Sustainability Communication

Jasmin Godemann • Gerd Michelsen
Editors

Sustainability Communication

Interdisciplinary Perspectives
and Theoretical Foundations

Editors
Dr. Jasmin Godemann
International Centre for Corporate Social
Responsibility (ICCSR)
Nottingham University Business School
Jubilee Campus, Wollaton Road
Nottingham, NG8 1BB, UK
jasmin.godemann@nottingham.ac.uk

Prof. Dr. Gerd Michelsen
Institute for Environmental and
Sustainability Communication (INFU)
Leuphana University
Scharnhorststraße 1
21335 Lüneburg, Germany
michelsen@uni.leuphana.de

ISBN 978-94-007-1696-4   e-ISBN 978-94-007-1697-1
DOI 10.1007/978-94-007-1697-1
Springer Dordrecht Heidelberg London New York

Library of Congress Control Number: 2011929937

© Springer Science+Business Media B.V. 2011
No part of this work may be reproduced, stored in a retrieval system, or transmitted in any form or by any means, electronic, mechanical, photocopying, microfilming, recording or otherwise, without written permission from the Publisher, with the exception of any material supplied specifically for the purpose of being entered and executed on a computer system, for exclusive use by the purchaser of the work.

Kind permission was given for use of parts of chapters originally published in 2005 by Oekom Verlag GmbH, München, Germany from "Handbuch Nachhaltigkeitskommunikation: Grundlagen und Praxis" in chapters 3, 5, 6, 7, 8, 10, and 16 in this volume.

Printed on acid-free paper

Springer is part of Springer Science+Business Media (www.springer.com)

Preface

Since the publication of the Brundtland Report in 1987 at the latest, there have been intensive discussions about the vision of 'sustainability' together with the related concept of 'sustainable development' in many different areas of society as well as in the scientific community. The degree of knowledge in the general population however is not very profound. At the same time it is argued that the concept of sustainable development can only be realised if there is broad support for its implementation in the general public. In order for this to happen it is necessary for much larger segments of society to become interested in this process and to become involved in this process. The pathway to the sustainable development of our society will only be taken when it becomes clear why the concept of sustainable development is a strategy for the survival of the human race.

Against this background there has been a growing awareness in recent years of the necessity of sustainability communication. This discipline has set itself the goal not only of providing a clear and persuasive understanding of sustainable development and of campaigning for its acceptance, but above all of involving people in the process of sustainable development and motivating them to actively take part in it. The scientific discourse accompanying this development is concerned with a number of different fields in sustainability communication and attempts to provide a theoretical foundation as well as a conceptual orientation for a communicatively based process shaping sustainable development.

This Handbook is meant as a contribution to that process, strengthening the theoretical grounding of sustainability communication and by using selected examples from such issues in sustainability as climate change or biodiversity showing which role sustainability communication can play in these fields. This involves learning to identify different levels and fields of sustainability communication but also in learning to recognise its limits. Sustainability communication cannot replace the decisions taken in politics and by individuals about possible courses of action, but it can accompany and support these processes. The Handbook should be seen as a compendium showing the spectrum of sustainability communication in all of its many facets, without however claiming to offer a complete review.

As is often the case, the writing and publication of this book involved many discussions, at times controversial, not only between the two editors but also with the authors. We have the distinct impression however that this volume has not only not suffered as a result, but it has on the contrary become better. We would like to take this opportunity to thank everyone involved for their cooperation, suggestions and criticism. We would also like to thank Paul Lauer for his patience and help in editing the language of the contributions. Even though everyone who took part in the writing of this Handbook was at pains to work carefully and precisely, there may still be errors in this publication. These are of course the sole responsibility of the editors.

Lüneburg and Nottingham   Gerd Michelsen
                                                                          Jasmin Godemann

Contents

Part I Mapping Sustainability Communication

1. **Sustainability Communication – An Introduction**............................. 3
   Jasmin Godemann and Gerd Michelsen

2. **Strong Sustainability as a Frame for Sustainability Communication**.. 13
   Konrad Ott, Barbara Muraca, and Christian Baatz

3. **Sustainability Communication: An Integrative Approach**.. 27
   Maik Adomßent and Jasmin Godemann

4. **Sustainable Communication as an Inter- and Transdisciplinary Discipline**... 39
   Jasmin Godemann

Part II Framework of Sustainability Communication

5. **Sociological Perspectives on Sustainability Communication**.............. 55
   Karl-Werner Brand

6. **Psychological Aspects of Sustainability Communication**................... 69
   Lenelis Kruse

7. **Media Theory and Sustainability Communication**............................. 79
   Claudia de Witt

8. **Communication Theory and Sustainability Discourse**....................... 89
   Andreas Ziemann

9. **Communicating Education for Sustainable Development**................. 97
   Inka Bormann

10	**Sustainability Communication:** **A Systemic-Constructivist Perspective**... Horst Siebert	109

Part III   Practice of Sustainability Communication

11	**Climate Change as an Element of Sustainability** **Communication** .... Jens Newig	119
12	**Biodiversity and Sustainability Communication** .... Maik Adomßent and Ute Stoltenberg	129
13	**Communicating Sustainable Consumption** .... Lucia A. Reisch and Sabine Bietz	141
14	**Corporate Sustainability Reporting**... Christian Herzig and Stefan Schaltegger	151
15	**Computer Support for Cooperative Sustainability** **Communication** .... Andreas Möller	171
16	**Participation: Empowerment for Sustainable Development** .... Harald Heinrichs	187
Index..		199

Contributors

Maik Adomßent is Senior Research Fellow in the Institute for Environmental and Sustainability Communication at the Leuphana University Lueneburg, Germany. His research interests include biodiversity and higher education for sustainable development.

Christian Baatz studied Environmental Sciences at the Leuphana University Lueneburg, Germany. He is a doctoral candidate at the University of Greifswald, Germany.

Sabine Bietz is a doctoral candidate at the Department of Consumer Behaviour and Consumer Policy at the University of Calw, Germany. Her main research interests are consumer behaviour, sustainable consumption and development.

Inka Bormann is Professor at the University of Marburg, Germany. Her main research areas are management instruments and their implications in the educational system and education for sustainable development.

Karl-Werner Brand is Professor of Sociology at the Technical University of Munich. From 1995–2005 he was director of the Munich Institute of Social and Sustainability Research (MPS), Germany. His key research activities are environmental sociology, environmental movements, ecological communication and sustainable consumption.

Claudia de Witt is Professor and member of the Institute of Educational Science and Media Research at the FernUniversität in Hagen, Germany, and she is chair of the Mobile Learning Group. Her research activities are focused on eLearning and mobile learning as well as theories of media education and media didactics.

Jasmin Godemann, Dr habil, is a Research Fellow at the Nottingham University Business School, UK. She was a Lecturer at the Institute for Environmental and Sustainability Communication, Leuphana University Lueneburg, Germany. Her main research interests include higher education for sustainable development, sustainability communication as well as inter- and transdisciplinarity in research and teaching.

Harald Heinrichs is Professor for Sustainability Policy at the Institute for Environmental and Sustainability Communication at the Leuphana University Lueneburg, Germany. His research interests include environmental and sustainability policy as well as media communication. He is also a consultant in the private sector.

Christian Herzig is Lecturer in Sustainability Accounting and Reporting within the International Centre for Corporate Social Responsibility at the Nottingham University Business School, UK. He was a Lecturer at the Leuphana University Lueneburg, Germany, and Visiting Research Fellow at the University of South Australia.

Lenelis Kruse is Professor Emeritus for Psychology at the German University for Distance Learning (Fernuniversitaet Hagen) until 2007 and is an honorary Professor at the University of Heidelberg, Germany. She was a member of the German Advisory Council on Global Change (WBGU) and is still active in the UN Decade on Education for Sustainable Development (2005–2014). Her main research interest is environmental psychology.

Gerd Michelsen is Professor at the Leuphana University Lueneburg and Director of the Institute of Environmental and Sustainability Communication, Germany. He holds the UNESCO Chair for Higher Education for Sustainable Development. His research interests include education for sustainable development and sustainability communication.

Andreas Möller is Professor at the Leuphana University Lueneburg, Germany, and member of the Institute for Environmental and Sustainability Communication. His research interests include corporate environmental informatics, new media for sustainability and the role of sustainability communication in organisations.

Barbara Muraca is Lecturer in Philosophy at the University of Greifswald, Germany. Her fields of research are process philosophy, environmental and sustainability ethics, feminist philosophy, and hermeneutics.

Jens Newig is Professor at the Leuphana University of Lueneburg, Germany and member of the Institute of Environmental and Sustainability Communication. His research centers around governance, participation, networks and media communication with a methodological focus on meta-analytical, experimental and transdisciplinary approaches.

Konrad Ott is Professor for Environmental Ethics at the Ernst Moritz Arndt University of Greifswald, Germany. His fields of research are discourse ethics, environmental ethics, theories of sustainability, justice and ethical aspects of climate change.

Lucia A. Reisch is Professor for Consumer Behaviour and European Consumer Policy at the Department of Intercultural Communication and Management at the Copenhagen Business School, Denmark. Trained as an economist and social scientist, she has been involved in research in sustainable consumption and production for two decades.

Stefan Schaltegger is Professor of Sustainability Management and Head of the Center for Sustainability Management at the Leuphana University Lueneburg. His research interests include corporate sustainability management, environmental and sustainability accounting, performance measurement and reporting and sustainable entrepreneurship.

Horst Siebert is Professor Emeritus for Adult Education at the Leibniz University Hannover. In 1970 he founded the first Institute for Adult Education in Western Germany. He is an Honorary Professor at the University of Jasi, Romania. His research interests are centred around teaching and learning in continuing education.

Ute Stoltenberg is Professor for Teacher Education/General Studies and Head of the Institute for Integrative Studies at the Leuphana University Lueneburg, Germany. Her research interests include biodiversity in education for sustainable development and the relationship between education and regional development in sustainable development.

Andreas Ziemann is Professor for Sociology of Media at the Bauhaus University in Weimar, Germany. His main research interests include communications and media studies, history and theory of modern societies, sociology of culture and the space of society.

List of Figures

Fig. 2.1	Theory of funds	22
Fig. 5.1	Dominant frames in the German sustainability discourse	60
Fig. 10.1	Constructivist disciplines	112
Fig. 10.2	Versions of constructivism	114
Fig. 11.1	Dynamics of news media communication about climate change	122
Fig. 12.1	Interactions between biodiversity, ecosystem services, human well-being, and drivers of change	131
Fig. 12.2	Ecological, social and socio-economic values of protected areas	137
Fig. 14.1	Perspectives of sustainable development and development of sustainability reporting	153
Fig. 14.2	Outside-in and inside-out approach to sustainability reporting	165
Fig. 15.1	Typical conversation	174
Fig. 15.2	Infrastructure of a web platform representing the user lifeworld	180

List of Tables

Table 2.1	Levels of the strong sustainability theory....................................	16
Table 4.1	Comparison of typology and previous categorizations.................	42
Table 9.1	Classification 'Gestaltungskompetenz' sub-competences............	104
Table 12.1	Demand for water in plant and animal production.......................	133

Part I
Mapping Sustainability Communication

Chapter 1
Sustainability Communication – An Introduction

Jasmin Godemann and Gerd Michelsen

Abstract The development of the term sustainability communication is accompanied by the call for responsible human interaction with the natural and social environment. This entails a process of social understanding that deals with the causes and with possible solutions. The task of sustainability communication is to critically evaluate and introduce an understanding of the human-environment relationship into social discourse. Alongside defining and providing a theoretical framework for mutual understanding, this chapter will describe issues, concepts and methods relating to sustainability communication.

Keywords Theoretical framework • Sustainability communication • Sustainability research • Systemic-constructivist perspective • Methods of sustainability communication

History of the Term

If the 1970s and the 1980s were above all characterised by debates about environmental problems, the 1990s were defined by political and economic discussions about so-called non-sustainable developments, globalisation and the concept of sustainable development. Triggered by the 1972 report of the Club of Rome,

J. Godemann (✉)
International Centre for Corporate Social Responsibility (ICCSR), Nottingham University Business School, Jubilee Campus, Wollaton Road, Nottingham, NG8 1BB, UK
e-mail: jasmin.godemann@nottingham.ac.uk

G. Michelsen
Institute for Environmental and Sustainability Communication (INFU), Leuphana University, Scharnhorststraße 1, 21335 Lüneburg, Germany

'The Limits to Growth', and followed by such landmark reports as 'Global 2000', published in 1980 by the Council on Environmental Quality, or the report 'Our Common Future' in 1987 by the Brundtland Commission, it became clear that humanity was entering a phase of radical social change calling for a new approach to dealing with anthropogenic environment problems, but also with improving humanity's ability to coexist in the world (Meadows et al. 1972; Council on Environmental Quality 1980).

This development is characterised by three closely woven basic trends. First, there is a rapid increase in global interrelationships in the *economy* through an ever greater flow of goods, money and information. Cheaper telecommunication and computer technologies on the one hand and denser and more closely linked networks of transport and energy supply on the other have changed global patterns of production, logistics and trade. Multinational corporations and transnational operating financial actors are attempting, and not without success, to influence these economic globalisation processes. On the other the globalisation of *ecological dangers* can be seen in anthropologic greenhouse effects, in climate change and in the loss of biodiversity. Global ecological dangers are linked with regional problems (such as water scarcity, flooding, forest damage, desertification, urban sprawl, famine, disease etc.) – and these in turn with local environmental damages (such as air pollution, waste, traffic noise, water pollution, losses in soil fertility, etc.). And thirdly the explosive increase in available information, the dissemination and the large-scale use of *modern information and communication technologies* have enabled the growth of data networks, and with it the expansion of research and development. What is in principle available worldwide however is not necessarily available locally. New and fast-growing inequalities in access to information, the so-called 'digital divide', deepen the divide between the winners and losers of global communication. All of these trends overlap, interlink and reinforce each other, thus leading to severe economic, ecological, social and cultural distortions both in individual regions as well as worldwide. The consequences of such developments worldwide can only be met if humans assume their responsibility and reshape their relationships to each other and the natural world. This requires a social process of mutual understanding that deals with both the causes of these developments and their possible solutions. In other words, a process of communication and mutual understanding that is also known as sustainability communication.

Sustainability and Communication

Before discussing sustainability communication, it is important to first clarify the contents and concept of sustainability and sustainable development. Sustainability has only recently found its way into academic discussions. At the latest since the 1992 United Nations Conference on Environment and Development in Rio de Janeiro and the final document 'Agenda 21' the concept of sustainable development has come to have a number of different interpretations and uses. Each social vision has a different weighting of the core elements of justice, a modest life, freedom and self-determination, participation, human well-being and responsibility for the future. Sustainability has been repeatedly held up as an important goal by governments, businesses, non-governmental organisations (NGO) and also at national and international conferences, playing a role in a variety of different interest groupings. Although the term is accompanied by imprecision, ambiguity and at times contradictions, there is a generally accepted understanding of what sustainable development means. The best formulation can be found in the report 'Our Common Future', also known as the Brundtland Report. "Sustainable development is development that meets the needs of the present without compromising the ability of future generations to meet their own needs" (WCED 1987: 43).

Sustainable development then is an ethically motivated normative concept referring to a form of economics and lifestyle that does not endanger our future. Such an ethical approach to shaping the future must ultimately be based on an understanding of strong sustainability (Daly 1997; Ott and Döring 2008), which in contrast to the concept of weak sustainability rejects the premise of unlimited substitutability of all natural resources with equivalents and considers this as irresponsible to future generations.

Science and research are thus challenged. So-called sustainability research is a paradigm shift within science. The focus is on the relationship between humans and the environment and the structure of research practice can be characterised as an integrated approach to cooperative problem-solving. Inter- and transdisciplinary research moves into the foreground, drawing attention to a new mode of knowledge production as well as a new understanding of science and confronting traditional scientific practice with a new mode of problem-oriented research that should give fundamentally different answers to the questions of today's complex society. The interests of social, economic and political actors are constitutive elements of the research process, expanding awareness of the problem and its potential solutions (Hirsch Hadorn et al. 2008).

Discussions about sustainable development are embedded in patterns of cultural perception and action (e.g. the issue of justice and equality). Research into mentalities and risk show that for example the perception of environmental phenomena as environmental problems depends on the cultural context, underlining the importance of cultural differences and their critical reflection. The vision of sustainability is also related to concepts of modernisation and development of society that entail a stronger engagement of individuals. Participation is often seen as a new challenge for political culture and thus has a close relationship to sustainable development. In this context communication can be understood as a social process in which common orientations are interchanged. "The necessity of communication can be found in the (anthropologic) circumstance that each consciousness is isolated, our neurophysiological, cognitive, emotional processes are mutually unobservable and there is no direct access to the thoughts, attitudes and intentions of the other. It is through communication that 'the interior is exteriorised', that we can inform each other, that we become social creatures. Communication is thus the principle of societal organisation itself" (Ziemann 2007: 124).

Sustainability communication is thus a process of mutual understanding dealing with the future development of society at the core of which is a vision of sustainability. It is both about values and norms such as inter- and intragenerational justice and about research into the causes and awareness of problems as well as about the individual and societal possibilities to take action and influence development. This process of mutual understanding takes place on a number of different levels and in different contexts: between individuals, between individuals and institutions, between institutions and within institutions, in schools and universities, in the media, in politics, in business, in communities and at regional, national and international levels. The success of communication about sustainability and a sustainable development depends then on a large number of factors, which does not simplify the process.

Essentially communication can be understood as symbolically mediated action, with humans constructing their reality on the basis of perceptions and experiences. This thesis is the foundation of much sociological thinking, whether Mead's (1934) symbolic interactionism or Berger and Luckmann's (1966) theory of the social construction of reality. The systems theoretical approach of Luhmann shows very nicely the value of communication: "Fish may die or human beings; swimming in lakes and rivers may cause illnesses; no more oil may come from the pumps; and average temperatures may rise or fall, but as long as this is not communicated it does not have any effect on society" (1986: 63). In summary it can be said that human behaviour, social values and attitudes towards the world and environment are mediated by communication.

The task of sustainability communication lies in introducing an understanding of the world, that is of the relationship between humans and their environment, into social discourse, developing a critical awareness of the problems about this relationship and then relating them to social values and norms. Scientific knowledge and scientific discourse play a central role in this undertaking to the extent that they contribute to strengthen or relativise the various positions and perspectives. Sustainability communication offers a framework for understanding a wide variety of social systems and actors (science, business, education, media, etc.).

Theoretical Framing of Sustainability Communication

In order to provide a theoretical framing for sustainability communication, a number of different scientific disciplines are needed, each with its own theoretical principles and knowledge. Theories that sustainability communication makes use of include systems theory and the epistemology of constructivism, approaches in media theory and in communication theory, as well as psychology and sociology. Sustainability communication still does not have its 'own' theoretical framework, such that one could speak of a theory of sustainability communication.

Given that sustainability communication is a process of the exchange of information between sender and receiver, an obvious starting point for the analysis and design of these exchange programmes would be to examine findings from communication theory. An important role here is played by common character encoding, language, values and norms, all of which allow the achievement of mutual understanding in communication and the establishment of stable social order. Sustainability communication is strongly influenced by mass media, which is needed to give it resonance, and therefore has a number of special characteristics:

- Reflexivity in regard to the problematic situation and how to handle it
- The establishment of sustainability as an intrinsic social value and the related issue of creating acceptance, with the possibility of different interpretations of sustainability clashing with each other
- The tendency to normalisation with the consequence that the more sustainability becomes a topic the less attention it receives and the less pressure there is to reach understanding
- And medialisation, which is an attempt to counter the tendency to normalisation in sustainability discourse by coupling it to the media

These characteristics have an impact on sustainability communication and should be taken into account when these processes are planned.

In this context findings from media theory research are important, in particular about the role of the media in disseminating an awareness of sustainability and influencing social discourse about sustainability. Social networks, made up of individual and group actors together with their 'ties', are becoming increasingly important. This is particularly the case in that mediated forms of communication (e.g. Web 2.0) and social interaction processes are part of these networks, with their own specific online use practices consisting of individual rules, network relationships as well as technical possibilities.

Another theoretical approach can be found in the systemic-constructivist perspective. Constructivism as a theory of perception and knowledge offers a way to explain the difficulty in communicating new ideas and knowledge to others. From a constructivist perspective learning is an intentional, self-controlled process. Research findings suggest that what is learned is not the same as what is taught, that individuals construct their own reality on the basis of previous experiences and come to their own understanding. New knowledge and new experiences have to 'fit' so that they are compatible with previous experiences and insights. This approach draws then attention to the importance of specific life experiences and of cultural and biographical differences.

Handling complexity and indeterminacy plays an essential role in the debate about sustainability and its core concept. There are still no conclusive answers as to how complex subject matter can be broken down so that the perception and analysis of problems become relevant for an individual. From a constructivist perspective the individual reduces the complexity of a subject step by step until he can integrate the new knowledge into his already existing stock of knowledge. In regards to action this means that if we want to confront reality critically then it is necessary to be able to first recognise and reflect on the our own perception as well as those of others. This also holds true for indeterminacy. There is no certainty in action. This uncertainty increases in the context of sustainability development and requires a critical awareness of risk as well as the ability to assess risk and a tolerance for ambiguity.

Sustainability always involves, either indirectly and directly, taking risk into consideration. There are numerous examples of risk including the risk of climate change, nuclear energy use, species loss, resource consumption, land use, noise pollution. Whoever would like to send a message is well advised to first understand how his communication partner perceives the world. Risk research shows that people deal with risk in a largely irrational manner (WBGU 1998). This is also an opportunity to approach risk from the perspective of natural and engineering sciences and communicate indicators and their critical values.

Comparative risk research reveals large differences in the perception of risk between different societies. This is known as the cultural relativity of risks, with the society in which one lives apparently determining which events are perceived as risks and feared. This is also valid for different lifestyle groups and milieus within a given society, with different forms of social organisation and lifestyles being associated with, for example, different images and understandings of nature and perceptions of danger. This thesis has received considerable support by comparative studies on risk perception (see for example Wildavsky 1993). Thus the statement that risk perception is culturally influenced does not simply mean that it varies from one country or culture to another, but that it also varies within a given country, with the social milieu in which an individual lives also helping to form his perception of risk. Taken together these two factors determine the importance of risk to an individual.

From a sociological perspective, the important question refers to the differentiation and change of lifestyles in the context of sustainability communication. Without using social science methods to conduct a detailed analysis of milieu and lifestyle together with the resulting consequences for 'marketing' the idea of sustainability, it would hardly be possible to develop a generalisable communication concept that would serve as an anchor for the idea of sustainability. The construct 'lifestyle' draws attention to the fact that with the increasing individualisation of society, the differentiation of economic conditions and educational biographies, with the varying use of mobility etc., a great number of different lifestyles have evolved. Lifestyles unite the use of resources, behaviours and value orientations to a pattern of life conduct. The development of different lifestyles is seen as an answer to individualisation in society, as the sociologist Ulrich Beck has described in his publications. Lifestyles are not emancipatory lifeplans but life patterns that are today closely related to patterns of consumption orientation. When sustainability communication is connected with changes in individual attitudes and behaviour, then differences in lifestyle take on a special importance.

A different context of sustainability communication is foregrounded in a further sociological perspective on the stabilisation and change of institutional practices through sustainability communication. This relationship between public communication and institutional change is a particularly important one to analyse. Especially noteworthy are the structuration theory of Anthony Giddens (1984), the symbolic interactionism of Peter L. Berger and Thomas Luckmann (1966) and the discourse-analytical perspective taken by Jürgen Habermas (1981).

Communication about sustainable development is also about communicating knowledge and stores of knowledge. However simply emphasising the meaningfulness of the concept of sustainability is not enough to mobilise change in a population. Environmental psychology suggests that the context of knowledge acquisition co-determines the relevance of knowledge for action (see Kruse in this volume). Knowledge needs a practical value and in order to understand sustainability different forms of knowledge are relevant. Expert subject knowledge alone is not enough. Systemic knowledge must be built up, that is knowledge of interrelationships, functions and processes. Only when one knows how to make use of this knowledge is there a capability to act. Systemic knowledge needs to be combined with the development of a system of values, with ethical orientations towards the relationship between humankind and nature, with direct experiences that involve emotionality and meaningfulness.

Methods in Sustainability Communication

A theoretical framework for sustainability communication is important in order to be able to understand the possibilities and conditions of communication processes about sustainability and its underlying concepts, to recognise its deficits and to analyse and develop it conceptually. However in order to be able to manage or influence the process of communication about sustainability, methods and instruments are necessary. These include for example social marketing, empowerment, instruments of participation and planning or education.

Social marketing is an important approach in sustainability communication and the same principles used in selling goods and services can be used to support a process of voluntary, individual behavioural change regarding such social issues as saving energy or conservation. The social marketing approach (Kotler and Lee 2008) provides a strategy for improving the efficiency of sustainability communication. This communication concept is oriented towards the needs of target groups and so towards lifestyles. Word-of-mouth communication is a central element of viral communication and today mainly takes place in online communication and in Web 2.0 social networks.

Another starting point for sustainability communication are empowerment strategies, which have as their goal to help people actively shape the conditions of their own life. This involves developing the competence to recognise non-sustainable activities and then apply knowledge about sustainability to remedy them. There is an institutional as well as an individual dimension to empowerment (see for example Wilkinson 1998). Communication and participation together with educational processes are meant to strengthen civil society, promote individual engagement and support political education processes that enable individuals to actively take part in shaping a sustainable society. A central role is played by increasing participation opportunities and the space for individuals to influence change in a sustainable way. This involves the ability to reflect critically on the uncertainties and risks, different types of rationality as well as the consequences of one's own actions, which are an intrinsic part of such an engagement. The use of a variety of different communicative planning and participation instruments plays a role here, from future workshops to future conferences as well as round tables and mediation or advocacy planning and eParticipation.

A broader context of sustainability communication involves examining education processes. Education has the medium and long-term goal of assisting learners to acquire the basic knowledge and competencies needed to actively shape a sustainable future for life and work as well as enabling them to participate and empowering them to take action. The goal of an education for sustainable development (ESD) is to help create the conditions for self-determined and autonomous action and not just to train changes in behaviour. ESD aims at developing and enhancing the creative potential in the individual, his competencies in communication and cooperative work as well as problem-solving and taking action. Learning processes need to be initiated that allow an individual to sharpen his awareness in both private and working life of what is ecologically responsible, economically feasible and socially acceptable as well as enabling him to make the corresponding changes in his behaviour. Such ESD processes take place in both the formal and informal educational sector.

Places and Contents of Sustainability Communication

Sustainability communication takes place on a number of different levels in the public sphere. The discussion involves arguments, possibilities to take action and positions on societal development, is derived from economic, ecological, social and cultural perspectives, and is found in a field of discourse that includes all social systems. This communication among different social systems, such as politics, law, science, business or education, works to prevent sustainability problems and their causes from being separated from economic or socio-cultural developments and encourages potential solutions to be examined in a holistic fashion.

Sustainability communication also addresses issues like biodiversity, climate, mobility or consumption, to name only a few examples. Two types of communication can be distinguished here. On the one hand there is societal discourse as communication about a specific topic and on the other there is the communication of a specific topic in order to achieve specific effects. Communication about a specific sustainability topic requires an inter- or transdisciplinary approach in order to comprehend both the breadth and the depth of a problem and its possible solutions. In communication of a specific sustainability topic the issue of communication methods and their effects is of greater interest.

When new topics, concepts and modes of sustainability communication are being developed and implemented it is important to also evaluate the specific measures and interventions so as to ensure and improve the quality of the concepts and programmes. The use of evaluation tools is however not always considered an integral part of sustainability communication. In sustainability communication, evaluation processes must also be adapted to the level action and the related communication processes.

In order to implement a vision of sustainability and of sustainable development, a diverse set of political instruments is needed. Since the concept of sustainable development involves not only the environmental idea but also a dimension of development, existing 'hard' and 'soft' environmental policy instruments relating to structural environmental policies need to be modified. Alongside the market and the state, civil society is an important instrument in achieving sustainable development goals.

Sustainability communication is classified as a 'soft' or persuasive instrument and is one of a number of information and advisory instruments that has gained popularity in the environmental policy field since the 1980s. Compared to regulatory and economic instruments (or so-called 'hard' instruments), 'soft' instruments have the great advantage that they are not subject to any special legal control or cumbersome coordination processes. For example, using community action to influence the behaviour of individuals can achieve considerable impact. At the same time involving citizens in the solution of their own problems opens up additional opportunities for influencing the future in a sustainable way.

References

Berger, P. L., & Luckmann, T. (1966). *The social construction of reality: A treatise in the sociology of knowledge.* Garden City, NY: Anchor Books.
Council on Environmental Quality (1980). *The global 2000 report to the president of the U.S.* (Vol. 2, the technical report). Washington, DC: U.S. Government Printing Office.
Daly, H. E. (1997). *Beyond growth: The economics of sustainable development.* Boston, MA: Beacon.
Giddens, A. (1984). *The constitution of society.* Berkeley: University of California Press.
Habermas, J. (1981). *Theorie des kommunikativen Handelns* (Vol. 1: Handlungsrationalität und gesellschaftliche Rationalisierung, Vol. 2: Zur Kritik der funktionalistischen Vernunft). Frankfurt am Main: Suhrkamp.
Hirsch Hadorn, G., Hoffmann-Riem, H., Biber-Klemm, S., Grossenbacher-Mansuy, W., Joye, D., Pohl, C., Wiesmann, U., & Zemp, E. (Eds.). (2008). *Handbook of transdisciplinary research.* Dordrecht: Springer.
Kotler, P., & Lee, N. (2008). *Social marketing: Influencing behaviors for good* (3rd ed.). Thousand Oaks: Sage.
Luhmann, N. (1986). *Ökologische Kommunikation. Kann die moderne Gesellschaft sich auf ökologische Gefährdungen einstellen?* Opladen: Leske Budrich.
Mead, G. H. (1934). In C. W. Morris (Ed.), *Mind, self, and society.* Chicago: University of Chicago Press.
Meadows, D., Meadow, D., Randers, J., & Behrens, W. (1972). *The limits to growth: A report for the Club of Rome's project on the predicament of mankind.* New York: Universe Books.
Ott, K., & Döring, R. (2008). *Theorie und Praxis starker Nachhaltigkeit.* Marburg: Metropolis.
WBGU (1998). Wissenschaftlicher Beirat der Bundesregierung Globale Umweltveränderungen: Strategie zur Bewältigung globaler Umweltrisiken. Jahresgutachten 1998. Berlin.
Wildavsky, A. (1993). Vergleichende Untersuchung zur Risikowahrnehmung: Ein Anfang. In B. Rück (Ed.), *Risiko ist ein Konstrukt* (pp. 191–211). Munich: Knesebeck.
Wilkinson, A. (1998). Empowerment: Theory and practice. *Personal Review, 27*(1), 40–56.
World Commission on Environment and Development (WCED). (1987). *Our common future.* Oxford: Oxford University Press.
Ziemann, A. (2007). Kommunikation der Nachhaltigkeit. Eine kommunikationstheoretische Fundierung. In G. Michelsen & J. Godemann (Eds.), *Handbuch Nachhaltigkeitskommunikation. Grundlagen und Praxis* (pp. 123–133). Munich: Oekom.

Chapter 2
Strong Sustainability as a Frame for Sustainability Communication

Konrad Ott, Barbara Muraca, and Christian Baatz

Abstract The term sustainability has enjoyed great success, but at the cost of overextending its meaning to the point of trivialization. There is such an overabundance of definitions, concepts, models and political strategies that it is not clear anymore whether the terms 'sustainability' and 'sustainable development' still bear any meaning. The theory outlined in this chapter counters these tendencies by identifying more precisely the normative field that constitutes the very core of the sustainability concept, while avoiding a too narrow understanding. It points out the ethical presuppositions as well as the requirements for a theoretical framework of a consistent and discursively justified concept of sustainability. This rectifies the vagueness of the term as currently used and offers new possibilities for sustainability communication.

Keywords Strong sustainability • Weak sustainability • Ethics • Philosophy • Natural capital

Understanding Sustainability

The complex idea of sustainability is the outcome of different intertwined threads running across history, societal movements, scientific research and political policy-making. After the Rio Summit, which contributed to establishing worldwide a discourse and communication framework for sustainable development, the term sustainability has often been used as a catchphrase without specific meaning. Some scholars consider the well-known definition of the Brundtland Report a bad

---

K. Ott (✉)
Environmental Ethics, Ernst Moritz Arndt University of Greifswald, Domstraße 11, 17487 Greifswald, Germany
e-mail: ott@uni-greifswald.de

B. Muraca • C. Baatz
University of Greifswald, Greifswald, Germany

compromise between the needs for nature conservation and aspirations for economic growth. While a broad framing of the sustainability concept allows for a diversified and wide-ranging participation of stakeholders in the implementation of sustainability, this vagueness also leaves it open to being misused by power groups who want to press their business-as-usual attitude into a new trendy setting, following the maxim 'If you can't beat them, join them!'

A more precise definition of the concept of sustainable development is needed, and one that offers a flexible and non-arbitrary orientation for action.

In the transdisciplinary field of sustainability discourse with its essentially communicative structure, the philosophical perspective has a number of important contributions to make. Crucial aspects of this contribution are:

- First, philosophy can play the role of a *mediator or messenger* by creating a bridge between the different 'voices' participating in the process – it can be a semantic bridge not only among different disciplinary languages, but also, and more especially, between non-formalized knowledge, intuitions, everyday assumptions as well as more formalized forms of knowledge (Muraca 2010). Moreover, philosophy can render accessible and subject to critique implicit intuitions about inter- and intragenerational justice, about duties towards the non-human world, about attributions of value emerging in different cultural and societal settings (economic, cultural valuation, livelihood values, preferences, spiritual and aesthetic valuations, etc.).
- Second, philosophy can play the role of the *gate-keeper* in discourse, by continuously verifying which voices have a stake and a place, who is permitted to talk and who is excluded from the communicative process. Moreover, philosophy has a critical role to play by making transparent the implicit and unquestioned assumptions behind arguments and demonstrating how powerful, mainstream lines of thought lead to the silencing of alternative perspectives on the question at issue (Muraca 2010).
- Furthermore, practical philosophy can act as a *participant* in discourse, rather than playing an observational role with regard to the different meanings, definitions and attributions of sustainability that are *factually* and often strategically employed in communicative processes within society. In this function philosophy introduces its own *methodologies and theoretical frameworks* into the communicative process.

This chapter focuses on this third role of practical philosophy, or more precisely, on how practical philosophy can frame the theoretical setting of sustainability discourse by developing a normative theory of sustainability, taking a clear stance in the scientific debate between weak and strong sustainability.[1] The theory

---

[1] In the international discourse on sustainability there are only a few approaches that attempt a philosophical and normative analysis from the point of view of inter- and intragenerational justice (see among others, Dobson 2003; Norton 2005). A thorough presentation of these approaches, involving a comparison with the theory of strong sustainability, would go beyond the scope of this chapter.

of strong sustainability presented in this chapter does not take as a mere given the pre-deliberative agreement on sustainability (as established after Rio in societal, political and scientific documents). This agreement combines commitments to future generations with the so-called three-pillar model, by which economic, environmental and societal objectives are to be (somehow) balanced. From a philosophical perspective, this is an insufficient foundation for a genuine discourse on sustainability. The theory of strong sustainability goes beyond this widespread agreement to critically address the very core of the sustainability idea (inter- and intragenerational justice, a diversified concept of 'natural capital' etc.) in order to shape a comprehensive normative theory that can offer a well-founded orientation to societal and political decision-making processes (Ott and Döring 2008; Grunwald 2009; Norton 2005).

Drawing on Habermas's discourse ethics, the theory of strong sustainability assumes that discourse is a particular form of communication in which argumentation takes place (Habermas 1981). Rather than being considered successful to the extent that actors achieve their individual goals, as is the case for strategic action, communicative action and its second-order mode of argumentation succeed insofar as the actors freely agree, on the basis of rationally supported arguments, that their goals are reasonable and acceptable by all participants. Thus in order to reconstruct the normative presuppositions that shape discourse one cannot simply, from a mere observational point of view, describe argumentation as it empirically and factually occurs; rather, from the participant perspective, it is possible to articulate the shared and often implicit ideals and rules that provide the reasons for regarding some arguments as better than others.

The theory of strong sustainability therefore aims at:

- identifying criteria for distinguishing sustainable and non-sustainable paths on the grounds of a wider consideration of arguments than merely economic ones,
- specifying the proper scope of the discourse by setting up a framework of fields of action and application,
- delivering a basis for operationalisation in policy and politics,
- performing as a 'rational corrective' to clarify the diffuse discourse on sustainable development taking place in society (Grunwald 2009).

By drawing on Lakatos's and Stegmüller's post-Popperian assumption that every theory is constituted by core elements and a set of applications, some of which are paradigmatic, some secure and some contested, the theory of strong sustainability avoids the risk of transforming sustainability into a 'theory about everything' without any specific boundaries of application. For example, global climate change would be a paradigmatic application of a theory of sustainability, whereas the issue of juvenile criminality in urban areas is only marginally related to sustainability issues, although not completely independent from them.

Consequently, the theory of strong sustainability consists of different 'levels' (see Table 2.1 below), which are not intended as a deductive hierarchy. The first two levels – the core elements of the theory – consist of a theoretical reflection framing the concept of sustainability as a regulative ideal. The last three levels open the field

Table 2.1 Levels of the strong sustainability theory (Ott and Voget 2007)

Level	Status in the theoretical framework
1. Idea (Theory of intra- and intergenerational justice)	Core of theory
2. Concept (*Strong* or *weak* sustainability, mediating concepts)	
3. Key principles (Resilience, sufficiency, efficiency)	Bridging principles
4. Fields of action (Nature conservation, agriculture and forestry, fisheries, climate change etc.)	Practical application
5. Target systems, specific concepts, indicators	
6. Implementation, institutionalisation, instrumentation	

for a fruitful exchange with policymaking, praxis and socially participatory actions. The third level aims at bridging theory and practice. By means of this structure it is also possible to identify different fields for communicative actions at different levels of the discourse.

Sustainability as an Ethical Concept

The 'ethics' of sustainability should not be equated with a comprehensive ethical theory (e.g. discourse ethics), a theory of justice (e.g. the theory of John Rawls 1973) or with environmental ethics. Instead, it presupposes that certain assumptions from discourse ethics, theories of justice and from the argumentative framework of environmental ethics can be used to elaborate the idea of sustainability (Ott 2004a).

The core of the idea of 'sustainability' consists in the issue of intra- and intergenerational distributive justice and encompasses duties *towards* currently living generations and future generations regarding different goods (see Norton 2005), with a special focus on natural resources (Ott and Döring 2008). The idea of sustainability thus links the obligatory dimension of moral reasons with a teleological perspective that takes different distributions of goods into account. Deontological obligations to posterity can be combined with an assessment of the consequences and side-effects of current actions and institutions in order to constitute a teleological perspective of how sustainable development might be established in policy- making. The deontological assumptions must be made explicitly. In terms of the responsibility of justice towards future generations, at a minimum the following questions must be addressed:

- Are there any obligations to future generations at all?
- Should responsibility for the future be based on an egalitarian-comparative standard or on an absolute standard?
- What can be considered a 'just' legacy?

When ethical questions of intergenerational duties are discussed, it has to be *first* justified whether there are any obligations to future generations at all (for a thorough analysis and refutation of so-called 'no obligation arguments', which deny the existence of such duties, see among others Ott 2004b). Neither Parfit's 'non-identity problem' nor the argument claiming that future persons cannot have rights today are convincing (Parfit 1987). In fact, they seem to contradict basic intuitions of duties towards future generations that most people across cultures and centuries have shared. Parfit's non-identity problem obtains its moral relevance by confusing the terms *individuality* and *personality* (Partridge 1990; Grey 1996; Ott 2004b). An argument against Parfit is that personality as a normative status is usually ascribed to human beings with specific cognitive capabilities. This status includes a system of rights. Individuality on the contrary refers to the concrete and contingent characteristics of a single human being resulting from a unique and non-interchangeable life story. Moral duties are applicable to a greater extent to personality than to individuality. Although the non-identity problem highlights the contingency involved on the level of individuality, its moral relevance regarding the justification of intergenerational duties is negligible. Accordingly, regardless of the specific individual identity that members of future generation might embody, they will still be 'persons' in the sense proposed here and therefore subjects of rights. Moreover, as Unnerstall has argued at length, future rights can justify present duties (Unnerstall 1999). The anticipatable impact of future (moral or juridical) rights of persons is a necessary and sufficient condition for current intergenerational duties with regards to different goods.

According to the *second* question, the ethical controversy centres on whether duties of justice towards future generations should be based on an *absolute* standard (access to anything that is required for a life of human dignity) or on a *comparative* one (no worse than current generations). The absolute standard ensures a 'basic human level' (in terms of basic capabilities, see below) whereas the comparative standard raises the issue of an appropriate 'equivalence'. While the former allows current generations to bequeath less to future ones than they themselves have inherited (provided that this would be sufficient to lead a decent or dignified human life), the latter requires that future persons be no worse off than current ones (on average). Many authors argue for a comparative standard. This also corresponds to widespread intuitions expressed in deliberative processes with stakeholders and practitioners. However, its ethical justification is in no way a trivial one as questions arise as to whether the approximate equality of intergenerational prospects of life should be aimed at for its own sake and whether it is morally relevant how spatially and temporarily separated groups of persons with different supplies of goods relate to each other.

The theory of strong sustainability argues on the one hand for a strong and demanding absolute standard and suggests replacing the 'basic needs' approach with a culturally interpretable and context-sensitive list of capabilities, such as compiled by Nussbaum (2001) in her 'broad and vague concept of the good' (Ott and Döring 2008). Whereas according to the basic needs approach all human beings are entitled to have merely what they need to survive, the capability approach sets the minimum standard at a much higher level so as to include all the necessary conditions to accomplish a good (rich, flourishing) life, i.e. a life worthy of a human being. This approach encompasses capabilities such as 'being able to live to the end of a human life of normal length; not dying prematurely, or before one's life is so reduced as to be not worth living'; 'being able to have attachments to things and people outside ourselves'; and 'being able to live with concern for and in relation to animals, plants and the world of nature'. The list is based on ideas of the intrinsic richness of human existence and on the idea that a good human life lies in the exercise and performance of specific human capabilities.

While anti-egalitarians deny that equality has any intrinsic value and thus limit intergenerational duties to an absolute standard (Frankfurt 1987), in the theory of strong sustainability also comparative aspects of justice above the absolute standard ought to be taken seriously. The comparative standard can be justified with the Rawlsian 'veil of ignorance' (Rawls 1973), which would have to be designed in such a way that the individuals behind it do not know to which generation they belong. Rawls's idea of *reciprocity*, which suggests an equal distribution as the starting point, leads to the conclusion that rational persons would probably choose a comparative standard as far as this is feasible within safe environmental limits.

The comparative standard can also be justified without recourse to Rawls. The conviction that from the moral point of view in the generational chain no generation is 'special' can be combined with a prohibition of primary discrimination (Tugendhat 1993) and the disputed 'presumption in favour of equality' (P). This constitutes a sufficient premise to shift the burden of proof in favour of an intergenerational comparative standard. The justification of P rests on the transfer of generally accepted principles (equal moral considerability of every person, equality before the law, equality of opportunity) to the sphere of distributive justice. In the end, both lines of justification converge to similar results.

The third core question leads to the next level of the theory, since it cannot be answered at the abstract level of theoretical moral justifications. It encompasses the widely debated issue about the 'fair bequest package' that current generations owe to future ones.

What Do We Owe to Future Generations? Arguments in Favour of 'Strong Sustainability'

Different approaches within communication about sustainability have to deal with the question at a conceptual level. A constitutive issue for the distinction between the various concepts of sustainability is the question of what legacy (the 'fair bequest package') current generations owe to subsequent ones. Legacies involve the production, preservation and reproduction of, in the language of economics, *packages of different kinds of capital*. The concepts of weak and strong sustainability diverge basically on what they respectively consider a fair bequest package. This is due to different assumptions regarding the extent to which natural capital can reasonably be *substituted* by human capital and man-made capital.

The concept of *weak sustainability* assumes far-reaching substitutability among different types of capital. Accordingly, a fair bequest package consists of a constant (cumulative) total level of capital. In practice, this means that nature can be consumed, provided that other capital reserves (man-made capital, human capital) are built up in its place. This would make it possible to envisage a future world where there were for example no forests, provided that all of the functions forests currently fulfil (production of wood, leisure functions, balancing effect on regional climate systems, etc.) can be satisfactorily fulfilled by artificial means (synthetic substances, nature films on TV, air conditioning etc.).

Weak sustainability envisages the different capital stocks of society in terms of an overall portfolio, in which natural capital is only one among a number of different stocks. The ideal portfolio manager would consider possibilities of substitution by trying to maximise the net present value. From this point of view the preservation of natural resources would be a meaningful and feasible goal only if it proved to be more efficient when compared to other income types. For the sake of comparability, natural resources have to be expressed in monetary terms. The deontological meaning of intergenerational duties can only be described in terms of a constraint imposed on maximization paths. The ethical idea is thus expressed as 'non-declining utility over time'.

Considering presumed limits of substitution between different capital stocks, advocates of *strong sustainability*, like Herman Daly (1997), plead for a diversely structured legacy. Regardless of the increase of other capital stocks, natural capital should be at least maintained at a constant level for the sake of future generations. Intuitively striking examples for the complementarity among capital stocks are the relations between fish and fishing boats, forests and lumber mills, crude oil and refineries etc. However, in principle this does not preclude the possibility of limited substitution in particular cases. For Daly (1997) the assumption of complementarity is a sufficient argument to justify the rule of strong sustainability, according to which natural capital should not decline over time (the constant natural capital rule – CNCR). However, further arguments can be introduced to justify the CNCR. In fact, it is not only about whether or not and to what extent nature *can* be substituted in the production process, but also about whether 'we' *would want* the ongoing substitution of nature with regard to the capabilities approach or, in other words, whether 'we' can justify this substitution in the eyes of future generations.

The concept of strong sustainability relies on a 'biospheric' *framing*: According to Daly (1997) the biosphere is characterised by living structures with a high degree of internal complexity, i.e. negentropic structures. The whole industrial economy is fundamentally reliant on the autopoietic regeneration of these very negentropic structures that, together with raw materials, constitute a specific type of capital, stocks and funds, which provides beneficial flows to human systems. Moreover, nature is not only seen as a repository of resources, but also as an interlinked ecological background in which economy and society are embedded.

The task of philosophical scrutiny is to develop a well-founded judgment that provides a guide to a reasoned choice between these two concepts of sustainability. The judging process takes place in due consideration of ethical principles and in a situation that is practice-oriented but without any direct pressure to act. It assumes the perspective of citizens as moral persons examining together reasons provided by theorists. The key arguments are (Ott and Döring 2008; Ott 2009):

- *Critique of the general economic framework on which the concept of weak sustainability relies*: A general reference to 'technological progress' or to economic models is not sufficient to justify weak sustainability. Such models are not at all neutral (Held and Nutzinger 2001); rather, if they make uncritical use of decisive economic concepts, such as substitution, discounting, and compensation, then they are part of the problem. An often given example of the falsification of the theory of weak sustainability is the insular state of Nauru in the Pacific Ocean (Gowdy and McDaniel 1999).
- *Multifunctionality of ecological systems*: A weighty argument against unlimited replaceability of natural assets is the multifunctional nature of many ecological systems. Specifically, for every single ecological function that a natural asset might possibly provide an artificial substitute must be identified. The substitutes must additionally be available now and not merely as a theoretical possibility. In addition, it is by no means certain that substitutes will always be of better value, have a lower risk or be more socially tolerable or 'prettier'.
- *Risk assessments and the precautionary principle*: In accordance with the precautionary principle, it would be wiser to opt for the concept of strong sustainability in case it turns out that after the consumption of large quantities of natural capital it proves to be indeed non-replaceable.
- *Greater freedom of choice for future generations*: It is by no means certain that people alive in the future will approve of current substitution processes. It does not necessarily follow from the fact that future preferences (beyond minimum requirements) are changeable that future generations will be delighted with a denatured, artificial world. The conservation of natural capital leaves more options open to people alive in the future.
- *Better compatibility with the argumentative framework of environmental ethics*: It is incontestable that strong sustainability pays greater respect to the diverse cultural, biophilic and spiritual values that people associate with the experience of nature and landscape. If, at the general level of environmental ethical discourses, people alive today speak, or learn to speak, authentically and autonomously about what natural assets and experiences of nature really mean to them, then they are thus (*ipso facto*) attempting to create an ethical tradition that should also be taught in environmental and nature education, should become habitual and should have some degree of permanence into the future. This leads to the question of which concept of sustainability best matches current insights, convictions and attitudes in the area of environmental ethics. Educationalists in the fields of environmental studies and nature conservation in particular could better convey the meaning and purpose of their activities within the framework of strong sustainability. Conversely, advocates of weak sustainability must – for conceptual reasons – regard current efforts in the field of nature education somewhat sceptically, even if they might not like to say so out loud to nature education practitioners.

These reasons can be considered sufficient to justify the concept of strong sustainability in an envisaged counter-factual discourse with representatives of future generations. Of course, neither Western ethicists nor economists are allowed to dictate a concept of sustainability to others; they may only raise it as a topic for discussion. However, it can be expected that the concept of weak sustainability, if its core premises are expatiated *coram publico*, might meet with surprise and refusal in many cultures.

Natural Capital

Some scholars oppose the term 'natural capital', arguing that nature should not be designated as a form of capital (Biesecker and Hofmeister 2009), arguing that the term capital tacitly implies transferring an understanding of utility resulting from the means of production, which is typical for man-made capital, to complex natural systems providing a variety of ecological services, whose components are living and subject to evolutionary alterations. In the theory of strong sustainability, 'capital' is used as a concept at the intersection of economics and philosophy, being neutrally defined as stocks yielding a somewhat beneficial or useful flow (Ott and Döring 2008). This concept of capital must be specified according to the specific features and benefits of different types of capital. Therefore, the theory of strong sustainability starts with the term *natural capital* in order to show in a subsequent step the 'differentiae specificae' of *natural* capital *as such*, especially the autopoietic productivity of the living.

Natural capital is a *totality* concept that encompasses heterogeneous entities. These entities can be described in terms of renewable and non-renewable stocks as well as living and non-living funds (Faber and Manstetten 1998). A *homogenised* understanding of natural capital contradicts the very meaning of the term. Single natural capital stocks are complex in themselves and, in addition, the actual components (soil, species, abiotic elements) are interlinked and interdependent (connectivity). Natural capitals are multiple, heterogeneous, and internally connected. The CNCR refers to this network of critical stocks. The definition of the term natural capital in the theory of strong sustainability is as follows: *natural capital consists of all components of animate and non-animate nature, especially living and non-living funds, that can benefit human beings and other highly developed animals in the exercise of their capabilities or that can constitute indirect functional or structural conditions for such beneficence in the broader sense* (Fig. 2.1).

Natural capital can be preserved by following 'management rules' as formulated by the German Advisory Council for the Environment (WBGU):

- Renewable resources may only be used at the rate at which they normally regenerate.
- Exhaustible raw materials and energy sources may only be consumed at the rate at which physically and functionally equivalent renewable substitutes are created.

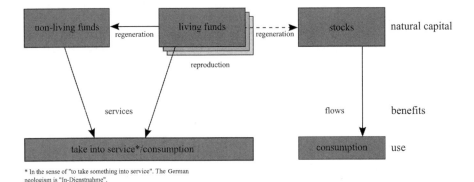

Fig. 2.1 Theory of funds (Source: von Egan-Krieger 2005)

- Pollutant emissions may not exceed the absorption capacity of environmental substances and ecosystems, and emissions of non-biodegradable pollutants are to be minimised, whatever the extent to which unoccupied storage capacity remains available.

The rule of preservation is to be understood as a prohibition of degradation and the rule of investment as a mandate for improvement and creative planning.

Conclusion

At first glance, a strategy for defining sustainability that is oriented to the factual use of words in everyday societal language seems most viable for the task of communicating about sustainability. However, as emphasised in the introduction to this chapter, this approach raises first and foremost the problem of unequal balances of power as well as the interest-influenced positioning that participants in communication processes are exposed to. Therefore, common sense and ordinary language should be taken as a point of departure for communication strategies but not as their final outcome. Instead, the theoretical concept formation proposed here is open to discursive intercourse on all levels, i.e. arguments can be examined, exchanged and improved.

Processes of reciprocal understanding about sustainability objectives and strategies belong to the category of 'epistemic-moral hybrids' (Potthast 2005) because they constitute an interface between science, ethics, economics and politics. An ethical perspective can provide, among other things, practical knowledge to guide action and provide some orientation for defining objectives. This knowledge is decisive for participatory decision-making processes. The theory of strong sustainability offers a feasible alternative to the popular three-pillar model, which has few proponents in academic discourse (Paech 2006). Moreover, the theory of strong sustainability can serve as a critical benchmark for a number of national and international activities such that their goals can no longer be determined arbitrarily. On the contrary, they should include programmes and strategies in those fields of action that are decisive for the preservation of critical natural capital.

With regard to sustainability communication the following consequences arise from the theory of strong sustainability. It is highly doubtful whether such concepts as the theory of strong sustainability, and related fields of discourse, will attract the attention of contemporary mass media in the short term. One should not however commit the fallacy of misplaced concreteness by identifying communication with mass media resonance. There are many platforms and arenas for communicating sustainability in the general and rational public sphere, in realms of civil society, academia, organisations and politics. A deeper understanding of the rational public sphere and its structures (Habermas 1992) might prevent communication strategies from promoting trivialization, 'anything-goes attitudes' or a 'race to the intellectual bottom'.

It is easy to argue that the regulative ideal of sustainable development is difficult to communicate since it is too vague, imprecise and cumbersome to be able to easily popularise. It is less easy to withstand this very danger. There are strong tendencies for the idea of sustainability to collapse into a platitude subsuming all possible (and impossible) sorts of issues under its umbrella. From a logical point of view, enlarging the scope of a concept comes at the price of a loss in meaning. Communicators should be aware of the logical relationship between scope ('extension') and meaning ('intension') threatening the meaning of sustainability. The theory of strong sustainability counters these tendencies by identifying more precisely the normative field constituting the very core of the sustainability concept, while avoiding a too narrow understanding; the theory of strong sustainability leaves the field open to and accessible for different perspectives, including intuitions, immediate experience, disciplinary approaches, non-formalized forms of knowledge and the like. Moreover, since the theory of strong sustainability does justice to the ecological view of the complexity of ecosystems and natural processes and allows for grounds for valuation and value considerations not restricted to a sheer economic view, it offers a range of arguments open to different actors in the field while also delivering a strong defensive ground against the risk of a colonization of our experiences and views by mainstream economic standards.

The theory of strong sustainability and sustainable development (a development that leads to sustainability as the ultimate regulative ideal) can be easily used to develop systems of objectives in fields corresponding to paradigmatic and even confirmed applications. This rectifies the vagueness of the term 'sustainable development' as currently used (and criticized) and offers new possibilities for sustainability communication. Communication strategies should take advantage of work by researchers and policy advisors to specify the concept of strong sustainability in different fields of environmental policy-making.

A thorough study of the sustainability debate and of the arguments delivered at different levels by researchers as well as by stakeholders plays a major role in the empowerment of citizens against manipulation by media and lobbies. A too vague and nebulous understanding of sustainability works to the advantage of those who have the factual power of establishing a definition of the term for decision-making, while discursive work on the concept based on a rational formulation of moral intuitions in the face of other real or virtual discourse participants challenges power-holders to deliver well-founded arguments and to reveal their assumptions and personal interests as well as the implications of their activities. Sustainability communication, which works through persuasion, can be considered an essential 'soft instrument' for implementing sustainability. However, it is important to note that persuasion can be accomplished in different ways. Following the theory of strong sustainability, the right balance for persuasive discourses is the so-called reflexive equilibrium between one's own basic intuitions and reasonable, rational arguments (including both common sense and a rational reflection transcending instrumental calculus), which take place in an intersubjective setting. Reflexive equilibrium requires a constant examination of one's own deepest beliefs in the face of the beliefs held by others, even when those others are not actually present. It enables a participatory process of learning and facilitates the further development and reinforcement of one's own ethical and social values while strengthening one's own sense of 'making a difference'. Lifestyles are also the outcome of habitualization processes, which can lead to a reduction of one's own options for social action. The reflexive process described here can thus have an emancipatory power and open to individuals new paths for the shaping of social patterns relevant for a sustainable development.

Promising ways to specify a concept are to make use of frames, images and visions, since they open the field for widely accessible 'story lines'. These seem necessary if the guiding principles and myths sustaining prevalent institutional practices that impede the diffusion of new concepts are to be challenged (see Brand in this volume). Sustainability will thus cease being a platitude and become a complex discursive field that provokes and even polarizes (see contributions in von Egan-Krieger et al. 2009).

References

Biesecker, A., & Hofmeister, S. (2009). Starke Nachhaltigkeit fordert eine Ökonomie der (Re) Produktivität. In T. von Egan-Krieger, J. Schultz, P. P. Thapa, & L. Voget (Eds.), *Die Greifswalder Theorie starker Nachhaltigkeit: Ausbau, Anwendung und Kritik* (pp. 169–192). Marburg: Metropolis.

Daly, H. E. (1997). *Beyond growth: The economics of sustainable development*. Boston, MA: Beacon.

Dobson, A. (2003). *Citizenship and the environment*. Oxford: Oxford University Press.

Faber, M., & Manstetten, R. (1998). Produktion, Konsum und Dienste in der Natur: Eine Theorie der Fonds. In F. Schweitzer & G. Silverberg (Eds.), *Selbstorganisation* (pp. 209–236). Berlin: Duncker & Humblot.

Frankfurt, H. (1987). Equality as a moral ideal. *Ethics, 98*, 21–43.

Gowdy, J., & McDaniel, C. (1999). The physical destruction of Nauru: An example of weak sustainability. *Land Economics, 75*, 333–338.

Grey, W. (1996). Possible persons and the problem of posterity. *Environmental Values, 5*, 161–179.

Grunwald, A. (2009). Konzepte nachhaltiger Entwicklung vergleichen – aber wie? Diskursebenen und Vergleichsmaßstäbe. In T. von Egan-Krieger, J. Schultz, P. Thapa, & L. Voget (Eds.), *Die Greifswalder Theorie starker Nachhaltigkeit. Ausbau, Anwendung und Kritik*. Marburg: Metropolis.

Habermas, J. (1981). *Theorie des kommunikativen Handelns*. Frankfurt/M: Suhrkamp.

Habermas, J. (1992). *Faktizität und Geltung*. Frankfurt/M: Suhrkamp.

Held, M., & Nutzinger, H. G. (Eds.). (2001). *Nachhaltiges Naturkapital: Ökonomik und zukunftsfähige Entwicklung*. Frankfurt/M/New York: Campus.

Muraca, B. (2010). *Denken im Grenzgebiet: prozessphilosophische Grundlagen einer Theorie starker Nachhaltigkeit*. Freiburg/München: Alber.

Norton, B. (2005). *Sustainability: A philosophy of adaptive ecosystem management*. Chicago: University of Chicago Press.

Nussbaum, M. (2001). *Women and human development: The capabilities approach*. Cambridge: Cambridge University Press.

Ott, K. (2004a). Noch einmal: Diskursethik. In N. Gottschalk-Mazouz (Ed.), *Perspektiven der Diskursethik* (pp. 143–173). Würzburg: Königshausen & Neumann.

Ott, K. (2004b). Essential components of future ethics. In R. Döring & M. Rühs (Eds.), *Ökonomische Rationalität und praktische Vernunft: Gerechtigkeit, ökologische Ökonomie und Naturschutz* (pp. 83–108). Würzburg: Königshausen & Neumann.

Ott, K. (2009). On substantiating the concept of strong sustainability. In R. Döring (Ed.), *Sustainability, natural capital and nature conservation* (pp. 49–72). Marburg: Metropolis.

Ott, K., & Döring, R. (2008). *Theorie und Praxis starker Nachhaltigkeit*. Marburg: Metropolis.

Ott, K. & Voget, L. (2007). Ethical dimension of education for sustainable development. *Education for Sustainable Development*, 1. Retrieved 18 Dec 2009, from http://www.bne-portal.de/coremedia/generator/pm/en/Issue__001/Downloads/01__Contributions/Ott__Voget.pdf.

Paech, N. (2006). Nachhaltigkeitsprinzipien jenseits des Drei-Säulen-Paradigmas. *Natur und Kultur, 7*, 42–62.

Parfit, D. (1987). *Reasons and persons*. Oxford: Clarendon.

Partridge, E. (1990). On the rights on future generations. In D. Scherer (Ed.), *Upstream/downstream: Issues in ethics* (pp. 40–66). Philadelphia: Temple University Press.

Potthast, T. (2005). Umweltforschung und das Problem epistemisch-moralischer Hybride. In S. Baumgärtner & C. Becker (Eds.), *Wissenschaftsphilosophie interdisziplinärer Umweltforschung* (pp. 87–100). Marburg: Metropolis.

Rawls, J. (1973). *A theory of justice*. Oxford: Oxford University Press.

Tugendhat, E. (1993). *Vorlesungen über Ethik*. Frankfurt/M: Suhrkamp.

Unnerstall, H. (1999). *Rechte zukünftiger Generationen*. Würzburg: Königshausen & Neumann.

von Egan-Krieger, T. (2005). Theorie der Nachhaltigkeit und die deutsche Waldwirtschaft der Zukunft. Diploma thesis, Greifswald University, Greifswald.

von Egan-Krieger, T., Schultz, J., Thapa, P. T., & Voget, L. (Eds.). (2009). *Die Greifswalder Theorie starker Nachhaltigkeit*. Marburg: Metropolis.

Chapter 3
Sustainability Communication: An Integrative Approach

Maik Adomßent and Jasmin Godemann

Abstract Sustainability communication is a relatively new concept. Its roots can be found in a number of different discourses, such as environmental, risk and science communication. On the one hand these discourses show a number of similarities, for example a similar thematic focus and the central role of the media. There are however clear differences concerning their theoretical foundations, political reach and respective actors. This contribution argues that sustainability communication should be seen as an integrative approach uniting the core elements of a number of different communication perspectives.

Keywords Environmental communication • Risk communication • Science communication • Sustainability communication • Integrative approach

An Outline of Sustainability Communication

Despite recent scientific findings on the global environment and the alarming reports that accompany them, sustainability does not seem to have become a near-term priority for society. In the process of changing this situation, an important role is given to sustainability communication. Its goal is to enable individuals and groups to develop the competences to adequately interpret the often contradictory and confusing scientific, technological and economic information available to them and then be able to react to and cope with the resulting long-term and complex societal challenges.

M. Adomßent (✉)
Institute for Environmental and Sustainability Communication, Leuphana University, Scharnhorststraße 1, 21335 Lüneburg, Germany
e-mail: adomssent@uni.leuphana.de

J. Godemann
International Centre for Corporate Social Responsibility (ICCSR), Nottingham University Business School, Jubilee Campus, Wollaton Road, Nottingham, NG8 1BB, UK

Considering the increasing relevance of sustainability communication, as for example debates of and reports on climate change in the media clearly show, it is necessary to find a theoretical foundation that would help locate sustainability communication, show its relationships to proximate discourses and specify its objectives. Sustainability discourse arises out of a number of different discourses, whose similarities include that they look back on relatively short histories, have been able to stimulate intensive discussions and will certainly also continue to do so in the future. The most important of these discourses are environmental communication, risk communication and science communication. Their different approaches are characterised by different foci, both at a theoretical and content level.

Environmental Communication

Environmental communication – and this is demonstrated by studies from Germany, Great Britain and other countries (BMU 2009; Defra 2008; Swanwick 2009) – has become a part of everyday communication. "Research and theory within the field are united by the topical focus on communication and human relations with the environment" (Milstein 2009). The discussion of various types of private, professional and social perception and the processing of complex environmental problems influences the public perception of the environment. "As we engage others in conversation, questioning, or debate, we translate our private concerns into public matters and thus create spheres of influence which affect how we and others view the environment and our relation to it" (Cox 2010: 26). Environmental communication includes every type of communication, whether delivered directly or by media, by individuals or institutions. This multi-facetted character of environmental communication makes it extremely difficult to find a unified definition. Within the scientific community it is also known as 'ecological discourse', with the sustainability concept being the most recent communicative 'framework'.

It was not until the beginning or the middle of the 1990s – or almost 10 years after the 'birth' of environmental communication in the United States with a publication 'Conservationism vs. Preservationism' in 1984 by Christine Oravec – that during a period of reflection following the earth summit meeting in Rio an awareness grew that the ideas from Agenda 21 had, in addition to their more global character, considerable importance for individuals (Oravec 1984). A decisive role in the gradual acceptance of the term environmental communication was surely also the cooperative potential in the concept of communication. Finally environmental communication is much more than just information or the transfer of knowledge. It is defined by neither consensus nor conflict. Instead it can be understood as a discursive place or possibility in which both poles can be formed (Coenen et al. 1998; Depoe et al. 2004). This potential to shape or optimise developments is a constitutive element of environmental communication, which is understood as a controllable process or single action resulting from an institution and addressed to either the population at large or a specific group of individuals.

In line with this broader understanding of the term, environmental communication is not simply a social phenomena to be observed. It can also be strategically influenced. This leads to the moral question about the ethical stance of the scientist. Every researcher must answer the question as to whether it is enough to work on analysing the relationship of humans to nature or whether he or she should use this knowledge to make a contribution to combating environmental abuses (see Peterson et al. 2007 on environmental communication as a 'crisis discipline').

Environmental Communication as a Key Instrument of Environmental Policy

Environmental communication goes beyond traditional ideas of communication as the dissemination of information, the findings of scientific research or the resulting policy options. Doubtlessly it is an important instrument of environmental policy, which however should not only be seen as governmental action. On the contrary, environmental communication can be understood as the sum of all efforts undertaken to develop society ecologically and sustainably (Hajer 1995). In societal communication the media play an important role not only for the creation of everyday knowledge but to a much greater extent for the transmission of information about global environmental changes. The medial handling of the environment is a difficult terrain. Due to their complexity, environmental topics are much more closely related than other issues to uncertainty and not-knowing and not-being-able-to-know. As a result uncertainty in society grows and worries about the future become more prevalent (BMU 2009). Mass media are most important for communication in modern societies because they are able to select and amplify the attention paid to a given topic and so influence public opinion (Maasen 2009).

In the classic understanding of environmental policy, environmental communication is commonly considered to be a persuasive (or informational or appellative) instrument. As a result its importance is often underestimated and it is classified as a 'soft' instrument, although it has a central function in terms of the implementation and acceptance of other instruments (Renn et al. 1995). It can be argued that environmental communication "can be better understood as a kind of 'basic instrument', namely in an adequate form it is necessary for the communicability and so the acceptability and functionality of all other instruments" (Mierheim 2002). Environmental communication should – especially when it is understood as 'communication for sustainability' – be considered a 'key concept'. "Environmental communication seeks to enhance the ability of society to respond appropriately to environmental signals relevant to the well-being of both human civilization and natural biological systems. [And that] scholars, teachers, and practitioners have a duty to educate, question, critically evaluate, or otherwise speak in appropriate forums when (…) communication practices are constrained or suborned for harmful or unsustainable policies toward human communities and the natural world" (Cox 2007: 15f).

Such optimism should be tempered with more sceptical assessments. At times there is even talk of a threatening 'crisis of environmental communication', whose cause however is not its lack of success but its insufficient self-reflection (Schack 2003). There should then be criticism of the criteria used to evaluate the success of environmental communication (as well as of those it applies to itself). Following Schack, the actual crisis of environmental communication is thus to be found "in its reaction to this description of crises with more and more activities before first clarifying its goals and requirements and especially its self-understanding" (Schack 2003: 162f). The basic orientations for actors in environmental communication (problem orientation, action orientation and/or empowerment orientation) represent constitutive elements, and ones that at the same time contain potential fault lines. Without greater transparency and reflection there is a danger that should these lines open up the reaction would largely be helpless (Schack 2003).

Risk Communication

As decisions have consequences that are not predictable, yet they are and must be taken, societal development is a process that is always accompanied by risks and the relationship of a society with its future changes as a result of the concern with risks. There are clearly discursive references between 'risk' and 'sustainable development'.

Risks can be divided into those in which human decisions and actions play a critical role in their origin, control or regulation and those that exist independently from human subjects and are neither attributable to nor justifiable by them. There is no 'objective' risk. Risk can be defined as a multi-dimensional construct, the individual or social creation of which involves many different aspects. Along with the perception, definition, calculation, assessment and regulation of the negatively experienced consequences of risk, there are also the calculable real negative consequences resulting from one's own decisions or another's, as well as dealing with existing risks (Beck 1992; Sellke and Renn 2010). Abstractly formulated, the question arises as to how on an individual as on a societal basis uncertainty can be dealt with in order to influence the future to one's own advantage. Certainty as well as health can be seen as 'concepts of reflection'. The opposite of these terms (uncertainty and illness) are reflected in them but are not realistic states and thus can never be actually achieved (Japp and Kusche 2008). Similarly the origin of a vision of sustainable development is tied to the communication of risk, since communication is essentially based on the perceived newness of the quality and dimension of risk, as expressed in such aspects as globality, complexity, extent and intensity of the damage potential, non-perceptibility, persistence and irreversibility as well as high conflict or mobilisation potential (WBGU 2000).

Theoretical presuppositions define the framework for the field of risk communication, which originally simply meant how well the public was informed about technological risks as sources of danger. This kind of risk communication takes place both preventively as well as in the event of actual damage. Risk communication in the classic sense consists of (as a rule scientific) experts educating laypeople, or 'normal' people, in order that they achieve some 'insight into necessity'. The meaning of this concept was expanded (similar to environmental communication) to a general term for the permanently re-occurring communication about health and environmental risks caused by humans – from printed warnings on cigarette boxes to feature-length television shows on global climate change (Doulton and Brown 2009; Lundgren and McMakin 2009; Sonnett 2009).

From a scientific perspective there has always been a close relationship between risk research and environmental research (Beck and Kropp 2007). Risk communication also has a political dimension – and so another similarity to environmental communication. In the final analysis environmental policy can be understood as risk management (Cox 2010). All measures in environmental policy are based on risk assessment or at least assumptions about risk – whether they are oriented toward the cooperation, precautionary or polluter-pays principle.

From a systems-theoretical perspective, society can be understood as a communication complex with a number of differentiated communicative contexts and so an equal number of different risks that are created by societal risk communication on a daily basis between such functional systems as politics and law, law and the economy and education and family. In such a perspective the societal dimension is however only one context that must be accounted for in risk communication. For a comprehensive understanding of risk communication in a 'poly-contextual risk society' the analytical framework needs to be expanded to handle both the problems of decision-making in a temporal dimension and the issues of coping with complexity in a factual dimension. Together these three horizons of meaning create the basic context of risk communication, which at the same time must always be communicatively realised (Japp and Kusche 2008). This perspective also eliminates the radical juxtaposition of experts and laypeople (Lorenzoni and Hulme 2009), replacing it with a recognition of a plurality of potentially complementary forms of knowledge. In addition technology and specialised knowledge are no longer considered a 'neutral' enclave of objectivity, making risk conflicts and discussions about the validity of scientific-technological knowledge the normal state (Juntti et al. 2009; McDonald 2009). Since mere information about risks diminishes the well-being of many individuals as well as their motivation to take action (Japp and Kusche 2008), risk communication now attempts to highlight less the dangers and more the opportunities to take action. If the opportunities to take action by participating actors are understood as resources, then such a resource communication approach can show how individual (personal resources) or collective (societal resources) competences to take action can be developed and/or made use of. This counters the fear of losing control in the face of environmental and health risks.

The media is, as with all of the approaches discussed here, of central importance (see Chap. 7). Media function as a very sensitive social alarm system or seismograph, registering tremors in the environment and in society. However "news [are] not an objective presentation of political reality, but an interpretation of events and issues from the perspective of reporters, editors and selected sources" (Wagner 2008: 27).

Risk communication cannot depend on the support of the media alone as they follow a different logic of action. "The media do not report on risks; they report on harms" (Singer and Endreny 1987: 10). However even if the political space of the global risk society is actually to be found in the media, those who feel obliged to report on environmental risks in a way that encourages action are confronted with journalistic selection and framing patterns of a sovereign and resistant system, which can hardly be changed in the short term.

Science Communication

Science communicates first and foremost with itself. The science system is characterised by self-referential unity, since in scientific work contexts knowledge is generated primarily in expert groups and mostly in a language that is incomprehensible to the non-scientific public. In addition the differentiation and increasing specialisation of the science system has led to each discipline developing, and continuing to develop, its own language. This barrier to understanding has led to an increasingly problematic boundary between science and the general public. The legitimacy of science, the quality of its achievements and its credibility are increasingly being criticised due to the ambivalence of new knowledge or the risks of technological developments and scientific research (e.g. genetic engineering). Since the middle of the twentieth century, the relationship between science and the public has been changed to the effect that, by the development and then pervasiveness of electronic media, a considerably larger 'mass democratic public' (Weingart 2003) has been established, which increasingly puts forward claims for greater participation in political processes, vocalises its interests and also attempts to realise them. Science is thus increasingly forced to open itself to and rethink its relationship with the general public. The public is becoming a relevant variable and the media has an important mediating function. The attempt to 'translate' and diffuse scientifically produced knowledge does not only take place across scientific boundaries but also within the system. Knowledge production is no longer a privilege of a special group of experts. Instead, it takes place in a number of different constellations of actors. In these inter- and transdisciplinary work contexts, not enough attention has been paid to the problem of translating and communicating this knowledge in a way that is adequate to its target groups (Wardekker et al. 2009).

The question arises as to the possibilities but also the limits to knowledge transfer, as well as the reasons for science 'turning to the public'. '*The public*' is an abstract and thus elusive concept and so the media takes its place as a representative. It takes on the function of assuring the selective attention of specific publics for science. In general there are three reasons for reporting on science (Göpfert and Peters 1996): (a) the utility argument, which is the concrete applicability and use value of information (e.g. specific health tips), (b) the culture argument, which views knowledge as an integral part of the creation of culture, (c) the democracy argument, according to which science and technology are of enormous importance for societal development and everyone must be informed in order to be able to take part in societal decision-making processes as responsible citizens. "The relation of science and society has undergone a few noticeable shifts over the past decades. All of these shifts are connected to a notion of democratisation" (Maasen 2009: 306).

The dissemination of scientific knowledge in public space is often described as a linear communication model, with a strict division between science and the public and with the public appearing as a passive and deficient addressee. Underlying this understanding of communication is a hierarchical model of forms of knowledge giving scientific knowledge a special position. Popularisation of scientific knowledge is then a 'top-down' process of education and reduced to a process of translation. "The deficiency model which claims that the public has a blank mind to be filled with scientific information and campaigns to promote the 'public understanding of science' as a means to obtain greater support and acceptability, simply have not worked to produce the desired outcome". A clear shift of emphasis needs to take place so that the public is recognised as democratic and actively expressing its interests and values concerning science (Maasen and Weingart 2005). Stehr speaks of the 'penetration of society with knowledge' (Stehr 1994), meaning that science is faced with a public that is itself increasingly scientifically trained. Continuing along the same lines, Felt (2002) argues for 'education through science' and criticizes the monopolistic position of scientific knowledge, demanding a 'new form of dialog culture'. In science communication the talk is of a 'dialogic turn', which is described "as a new form of scientific governance based on dialogue, interaction and participation throughout the research process rather than the unidirectional knowledge transfer of completed research results from researchers to policy-makers, practitioners and members of the public" (Phillips 2009).

This shifting of perspective, i.e. the 'disenchantment' of the special position of scientific knowledge, allows a new view of communication between science, the media and other social functional areas. Because of the question about the perception and selection of different systems within the communication process, the discourse on scientific topics within this troika are not only interesting from a scientific or media-sociological perspective. Against this background there would actually be, according to Fischhoff (2007: 5), a need for more experts in order to have effective science communication. "Creating scientifically sound communication requires recruiting and coordinating three kinds of experts: domain scientists, to represent the research about climate change and its effects; decision scientists, to identify the information critical to specific choices; and social scientists, to identify barriers to communicating that information and to create and evaluate attempts to overcome those barriers. It also requires designers, to implement communication concepts in sustainable ways".

In the context of sustainability communication this new emphasis on the relationship between science and the public offers new insights. What role does sustainability communication play in this mechanism of knowledge transfer? Does sustainability communication create transition channels between these systems so that knowledge diffuses between the public and science, leading to clear decisions in the political system? Measures to achieve a sustainable development must be accepted and supported by society. This requires a public awareness of the problem, which however should not be stirred up by an alarmism as often takes place in medial environmental reporting. Sustainability requires that short-term thinking be replaced by long-term thinking: "long-term demands for coherence on the basis of the sustainability postulate seem however (...) not to be compatible with the mechanism of an alarmed problem awareness" (Grunwald 2004). Science is given the role of acting in a critical fashion towards the public's awareness of problems, i.e. either sensitising itself for certain problems or relativising already established problems and possibly modifying them. This function is a central interface of science to society. In the context of sustainability science, communication can at this point be supplemented by sustainability communication, which knows the selection criteria and communication structures of the media system but does not make use of this alarmism itself to create attention. Sustainability communication has the role of sensitising a scientifically generated awareness of problems to questions of sustainable development and introducing them adequately into the public discussion.

Framing Sustainability Communication

This comparative assessment of environmental communication, risk communication and science communication shows that there are a number of similarities that are also constitutive for sustainability communication. All areas show a large number of commonalities with the discipline of communication sciences, while at the same time they are metadisciplinary fields of research that cross other scientific disciplines.

All discourses are united by a topical focus, which (especially for environmental and risk communication) are mainly directed at environmental and/or health relevant issues. These are largely characterised by a high degree of complexity, which given the reliability of scientific knowledge is always connected with a certain degree of uncertainty. Accordingly target group specific communication about uncertainty plays a central role – whether political decision-makers are being addressed or complex factual matters are being presented in the mass media (Kloprogge et al. 2007; Wardekker et al. 2009).

Furthermore all these strands have changed from a passive (self-) understanding (communication about…) to an active intervention (communication for…) (Moser and Dilling 2008). Instead of the educational transmission of information, the focus will always be more on aspects of pluralisation and the participation of affected and potentially affected individuals. It is noticeable how particularly for risk communication there has been a change from a corrective orientation to a preventative approach, as has already taken place in other discourses. In this context it is only consistent when the role of the media is considered as central across all disciplines – especially regarding their function (social seismograph versus controlling authority) – but can also be controversially discussed. The theoretical foundation of these discourse strands has advanced in varying degrees. The most advanced is certainly risk communication, which environmental communication partially refers to. The latter is a special case to the extent that it is best realised or can be best realised in sustainability communication and is an essential – of the discourses discussed above perhaps the most important – building block of this integrative approach. Finally all three approaches have in common a factual, a social and a temporal dimension and extend over a sphere of action that can reach from the local to the global dimension.

These claims can be specified using the example of climate change, because much of what is known or assumed about climate change communication is inferred from studies in other fields (e.g., risk communication, science communication, (mass) media communication, social marketing or rhetoric). "Challenges that communicators face in trying to convey the issue are somewhat typical for many sustainability-related topics, as they encompass characteristics like invisibility of causes, distant impacts, lack of immediacy and direct experience of the impacts, lack of gratification for taking mitigative actions, disbelief in humanity's global influence, complexity and uncertainty, inadequate signals indicating the need for change, perceptual limits and self-interest" (Moser 2010: 31).

Since media has no 'magic bullet' for informing the public, communication designers have to make their best possible efforts to identify the information most worth knowing and focus their communication outreach accordingly (Maibach and Hornig Priest 2009). A constantly growing body of research explores what kind of information is effective in influencing the public's perception of climate change, concluding that information should always be tailored to different public groups according to their beliefs and attitudes. There is evidence that effective scenarios might help people to relate to climate change, given that impacts can be presented both for the near future and the longer term, and for socio-economic changes in their local region (Lorenzoni and Hulme 2009; Ereaut and Segnit 2006; Segnit and Ereaut 2007). Visualisation of abstract phenomena might also be helpful. But care should be taken in using frightening images because although they may initially attract public attention, they are also likely to disempower individuals, distancing them from the issue. As O'Neill et al. and Nicolson-Cole (2009) state, it is more fruitful to use, in combination with dramatic images, 'enabling' images that the target audience can relate to.

With regard to sustainability communication, as exemplified by the 'Boulder Manifesto' for the field of climate change communication (Harriss 2008), the bottom line should be a kind of resource communication that keeps in mind that, together with natural and economic resources, people's knowledges, abilities and skills are the most important resources for change.

References

Beck, U. (1992). *World risk society: Towards a new modernity*. London: Sage.
Beck, U., & Kropp, C. (2007). Environmental risks and public perceptions. In J. Pretty, A. S. Ball, T. Benton, J. Guivant, D. Lee, D. Orr, M. Pfeffer, & H. Ward (Eds.), *Handbook on environment and society* (pp. 601–612). Los Angeles/London: Sage.

BMU – Federal Ministry for the Environment, Nature Conservation and Nuclear Safety (Ed.) (2009). *Environmental awareness and sustainable consumption.* Retrieved July 30, 2010, from www.umweltbundesamt.de/umweltbewusstsein-e/umweltbewusstsein.htm.
Coenen, F., Huitema, D., & O'Toole, L. J., Jr. (Eds.). (1998). *Participation and the quality of environmental decision making* (Environment & Policy, Vol. 14). New York: Academic.
Cox, R. (2007). Nature's "crisis disciplines": Does environmental communication have an ethical duty? *Environmental Communication: A Journal of Nature and Culture, 1*(1), 5–20.
Cox, R. (2010). *Environmental communication and the public sphere* (2nd ed.). Thousand Oaks: Sage.
Defra (Department for Environment, Food and Rural Affairs) (2008). *A framework for pro-environmental behaviours.* Report. London.
Depoe, S. P., Delicath, J., & Aepli Elsenbeer, M. (Eds.). (2004). *Communication and public participation in environmental decision making.* New York: State University of New York Press.
Doulton, H., & Brown, K. (2009). Ten years to prevent catastrophe? Discourses of climate change and international development in the UK press. *Global Environmental Change, 19*, 191–202.
Ereaut, G. & Segnit, N. (2006). Warm words. *How are we telling the climate story and can we tell it better?* (Institute for Public Policy Research) Retrieved July, 30, 2010, from www.ippr.org/publicationsandreports.
Fischhoff, B. (2007). Nonpersuasive communication about matters of greatest urgency: Climate change. *Environmental Science & Technology, 41*(21), 7204–7208.
Felt, U. (2002). Bildung durch Wissenschaft. *DIE – Zeitschrift für Erwachsenenbildung, 9*(2), 22–25.
Göpfert, W., & Peters, P. (1996). Wissenschaftler und Journalisten – ein spannungsreiches Verhältnis. In W. Göpfert & S. Ruß-Mohl (Eds.), *Wissenschaftsjournalismus: Ein Handbuch für Ausbildung und Praxis* (3rd ed., pp. 21–27). Berlin: Econ.
Grunwald, A. (2004). Die gesellschaftliche Wahrnehmung von Nachhaltigkeitsproblemen und die Rolle der Wissenschaften. In D. Ipsen & J. C. Schmidt (Eds.), *Dynamiken der Nachhaltigkeit* (pp. 313–341). Marburg: Metropolis.
Hajer, M. A. (1995). *The politics of environmental discourse: Ecological modernization and the policy process.* Oxford: Oxford University Press.
Harriss, R. (2008). An ongoing dialogue on climate change: The Boulder Manifesto. In S. C. Moser & L. Dilling (Eds.), *Creating a climate for change: Communicating climate change and facilitating social change* (pp. 485–490). Cambridge: Cambridge University Press.
Japp, K., & Kusche, I. (2008). Systems theory and risk. In J. O. Zinn (Ed.), *Social theories of risk and uncertainty: An introduction* (pp. 76–105). Malden: Wiley Blackwell.
Juntti, M., Russel, D., & Turnpenny, J. (2009). Evidence, politics and power in public policy for the environment. *Environmental Science & Policy, 12*, 207–215.
Kloprogge, P., van der Sluijs, J. P., & Wardekker, J. A. (2007). *Uncertainty communication: Issues and good practice.* Utrecht: Copernicus Institute for Sustainable Development and Innovation, Utrecht University.
Lorenzoni, I., & Hulme, M. (2009). Believing is seeing: Laypeople's views of future socio-economic and climate change in England and in Italy. *Public Understanding of Science, 18*, 383–400.
Lundgren, R. E., & McMakin, A. H. (2009). *Risk communication: A handbook for communicating environmental, safety, and health risks* (4th ed.). Hoboken: Wiley.
Maasen, S. (2009). Converging technologies – Diverging reflexivities? Intellectual work in knowledge-risk-media-audit societies. In M. Kaiser, M. Monika Kurath, & S. Maasen (Eds.), *Governing future technologies, sociology of the sciences* (Yearbook, Vol. 27). Dordrecht: Springer.
Maasen, S., & Weingart, P. (Eds.). (2005). *Democratization of expertise? Exploring novel forms of scientific advice in political decision-making* (Sociology of the Sciences, Vol. 24). Dordrecht: Springer.
Maibach, E., & Hornig Priest, S. (2009). No more "business as usual" addressing climate change through constructive engagement. *Science Communication, 30*(3), 299–304.
McDonald, S. (2009). Changing climate, changing minds: Applying the literature on media effects, public opinion, and the issue-attention cycle to increase public understanding of climate change. *International Journal of Sustainability Communication, 4*, 45–63.

Mierheim, H. (2002). Der neue Stellenwert der Umweltkommunikation in der Umweltpolitik. In Umweltbundesamt (UBA) (Eds.). Perspektiven für die Verankerung des Nachhaltigkeitsleitbildes in der Umweltkommunikation – Chancen, Barrieren und Potenziale der Sozialwissenschaften (pp. 1–2). Berlin: Erich Schmidt.

Milstein, T. (2009). Environmental communication theories. In S. W. Littlejohn & K. A. Foss (Eds.), *Encyclopedia of communication theory* (pp. 344–349). Thousand Oaks: Sage.

Moser, S. C. (2010). Communicating climate change: History, challenges, process and future directions. *Wiley Interdisciplinary Reviews (WIREs) Climate Change, 1*(1), 31–53.

Moser, S. C., & Dilling, L. (2008). Toward the social tipping point: Creating a climate for change. In S. C. Moser & L. Dilling (Eds.), *Creating a climate for change: Communicating climate change and facilitating social change* (pp. 491–516). Cambridge: Cambridge University Press.

O'Neill, S., & Nicholson-Cole, S. (2009). "Fear won't do it": Promoting positive engagement with climate change through visual and iconic representations. *Science Communication, 30*(3), 355–379.

Oravec, C. L. (1984). Conservationism vs. preservationism: The "public interest" in the Hetch Hetchy controversy. *Quarterly Journal of Speech, 70*, 444–458.

Peterson, M. N., Peterson, M. J., & Peterson, T. R. (2007). Environmental communication: Why this crisis discipline should facilitate environmental democracy. *Environmental Communication: A Journal of Nature and Culture, 1*, 74–86.

Phillips, L. J. (2009). Analyzing the dialogic turn in the communication of research-based knowledge: An exploration of the tensions in collaborative research. *Public Understanding of Science*, OnlineFirst, published on 18 Aug 2009, 1–21.

Renn, O., Webler, T., & Wiedemann, P. (Eds.). (1995). *Fairness and competence in citizen participation: Evaluating models for environmental discourse* (Risk, Governance and Society, Vol. 10). Dordrecht: Springer.

Schack, K. (2003). *Umweltkommunikation als Theorielandschaft: Eine qualitative Studie über Grundorientierungen, Differenzen und Theoriebezüge der Umweltkommunikation*. München: Oekom.

Segnit, N. & Ereaut, G. (2007). *Warm Words II: How the climate story is evolving and the lessons we can learn for encouraging public action*. London. Retrieved July 30, 2010, from www.ippr.org/publicationsandreports.

Sellke, P., & Renn, O. (2010). Risk, society and environmental policy: Risk governance in a complex world. In M. Gross & H. Heinrichs (Eds.), *Environmental sociology: European perspectives and interdisciplinary challenges* (pp. 295–322). Dordrecht: Springer.

Singer, E., & Endredy, P. (1987). Reporting hazards: Their benefits and costs. *Journal of Communication, 37*(3), 10–26.

Sonnett, J. (2009). Climates of risk: A field analysis of global climate change in US media discourse, 1997–2004. *Public Understanding of Science*, OnlineFirst, published on 9 Oct 2009, 1–19.

Stehr, N. (1994). *Knowledge societies*. London: Sage.

Swanwick, C. (2009). Society's attitudes to and preferences for land and landscape. *Landscape and Urban Planning, 59*(1), 1–11.

Wagner, T. (2008). Reframing ecotage as ecoterrorism: News and discourse of fear. *Environmental Communication: A Journal of Nature and Culture, 2*(1), 25–39.

Wardekker, J. A., van der Sluijs, J. P., Janssen, P. H. M., Kloprogge, P., & Petersen, A. C. (2009). Uncertainty communication in environmental assessments: Views from the Dutch science-policy interface. *Environmental Science & Policy, 11*, 627–641.

WBGU (German Advisory Council on Global Change). (2000). *World in transition: Strategies for managing global environmental risks. Annual Report 1998*. Berlin: Springer.

Weingart, P. (2003). *Wissenschaftssoziologie*. Bielefeld: Transcript.

Chapter 4
Sustainable Communication as an Inter- and Transdisciplinary Discipline

Jasmin Godemann

Abstract The goal of sustainability communication is to generate knowledge in inter- or transdisciplinary research processes and then have it enter the public discussion. The following chapter discusses the basic terms 'interdisciplinarity' and 'transdisciplinarity' and reflects upon the results of collaboration within heterogeneous groups. The challenge is that such communication processes must achieve understanding between individuals who – in regards to the object of communication – have systematically different scientific perspectives and everyday points of view. Finally inter- and transdisciplinary collaboration is characterised as social learning and an argument is made for the creation of frameworks that enable such a form of collaboration.

Keywords Interdisciplinarity • Transdisciplinarity • Sustainable science • Group factors • Collaboration • Knowledge integration

Sustainability Research and Communication

Sustainability communication moves in a special network of relationships among the three spheres of science, the public and practice. It is concerned with questions and problems that can be characterised as so-called 'in-between' phenomena, the analysis of which involves intensive cooperation among scientists of different disciplines together with representatives of societal praxis. "And if these problems refuse us the favour of posing themselves in terms of fields or disciplines, they will demand

J. Godemann (✉)
International Centre for Corporate Social Responsibility (ICCSR), Nottingham University Business School, Jubilee Campus, Wollaton Road, Nottingham, NG8 1BB, UK
e-mail: jasmin.godemann@nottingham.ac.uk

of us efforts going as a rule beyond the latter" (Mittelstraß 2002: 2). Sustainability communication can be seen as part of the larger field of sustainability science, which itself can be understood as a change of perspective within the scientific landscape. It is focused on the human-environment relationship; the structure of research practice can be characterised as an integrated approach to cooperative problem-solving. Questions of sustainable development comprise a number of sub-problems that are typically addressed by different disciplines. An important task of sustainability communication is to make the knowledge – together with a sensitivity towards these problems – that is created in these often inter- and transdisciplinary research processes available to the public for discussion. The communication processes involved cross boundaries by overcoming both disciplinary and scientific boundaries. Indeed for inter- and transdisciplinary sustainability research it can be said that "without (…) successful communication, the research simply does not happen" (Nilles 1975: 12) and for cross-border sustainability communication it is equally true that without adequate communication sustainability will not gain entry into society. Communication in inter- and transdisciplinary teams in sustainability research is discussed below. First of all, the two central terms – interdisciplinarity and transdisciplinarity – will be explained.

The Terms Interdisciplinarity and Transdisciplinarity

Interdisciplinary Collaboration

The term 'interdisciplinarity' has a long history and it is broadly discussed as a concept, a methodology, a process, a way of knowing or even a philosophy (OECD 1972, 1998; Thompson Klein 1990, 1996, 2010; Weingart and Stehr 2000; Lattuca 2001, 2002; Mittelstraß 2002; Aram 2004; Derry et al. 2005; Aboelela et al. 2007; Thompson Klein et al. 2010). All interdisciplinary activities – whether in research or in teaching – have in common the fact that they are rooted in the idea of constructing a comprehensive understanding and synthesis of knowledge. Interdisciplinary includes the attempt to integrate various insights into some sort of coherent overall concept, theme or metaphor. "Interdisciplinarity, rather, has to begin at home, in one's own mind. It is connected with an ability to think 'laterally', to question what others have not questioned, to learn what is not known within one's own discipline" (Mittelstraß 2001: 397). Interdisciplinarity as a form of cross-border, coordinated collaboration between different scientific disciplines means, first, that interdisciplinarity is pursued in a *coordinated* fashion and needs someone who in addition to his or her specialized work takes on the tasks of coordinating work. Second, interdisciplinarity as a form of *collaboration* subordinates the various perspectives to one research interest and the various methods to one research goal. This involves integrating different perspectives and skills from the disciplines involved at different phases of the process. Third, there are different versions of interdisciplinarity.

Weaker versions include the exchange of ideas and opinions, literature reception and the use of sources outside one's own discipline, while stronger versions involve working together on a common problem. And, fourth, interdisciplinarity serves the research goal, and a goal should attain results that a single disciplinary approach alone could not achieve.

The latter aspect of viewing a problem across disciplinary borders and broadening perspectives shows clearly that interdisciplinarity must be judged by the level of integration of knowledge it achieves. In the literature there are a number of different models that can be classified along a continuum from less to greater knowledge integration. Lattuca (2001: 112ff.) describes four levels of interdisciplinarity (a) as *informed disciplinarity*, in the sense that disciplines borrow methods and instruments from each other. This form of interdisciplinarity cannot ultimately be classified as such since other disciplinary perspectives are not integrated but merely adapted for use (b) as *synthetic interdisciplinarity*, which implies a closer relationship of disciplines with the research question being considered from different disciplines. This model however also does not involve an integration of disciplinary perspectives. Questions affecting more than one discipline are discussed, but the perspectives on the problem are additive, aligned side by side (c) as *transdisciplinarity*, understood as a principle that is beyond any disciplinary borders, existing above the problem as a unified worldview. The goal is to unite knowledge and develop a generalized and axiomatic transcendence of disciplines (d) as *conceptual interdisciplinarity*, which focuses on the problem and makes use of a number of disciplines to contribute to a solution. This involves critical reflection on and integration of disciplinary knowledge

Lattuca has systematised the prevailing terminology in the literature (for example, Jantsch 1972; Kockelmans 1979; Thompson Klein 1990) and ordered them in relationship to their typology. This typology is especially helpful for the evaluation of inter- or transdisciplinary collaboration. It shows that the term covers different levels of disciplinary perspective integration and that these have a corresponding effect on the results or product of research.

Transdisciplinary Collaboration

Within the context of sustainability research, the term has been used in a still broader sense than described by Lattuca (Table 4.1) (Hirsch Hadorn et al. 2008). Interdisciplinarity is an approach that transcends the boundaries of a segmented thinking within science. The transdisciplinary approach also involves a non-scientific perspective: "Transdisciplinarity moves beyond 'interdisciplinary' combinations of academic disciplines to a new understanding of the relationship of science and society" (Thompson Klein 2004: 517). The term transdisciplinarity became popular during the mid-1990s in the discussion about a new type of knowledge production and a new understanding of science. Traditional scientific practice (mode 1) is confronted with a new mode of research and it becomes clear that the new problem-related mode of

Table 4.1 Comparison of typology and previous categorizations from Lattuca (2001: 114)

Informed disciplinarity	Instrumental interdisciplinarity
	Pseudo-interdisciplinarity
	Cross-disciplinarity
	Partial interdisciplinarity
Synthetic interdisciplinarity	Instrumental or cross-disciplinarity (motivated by an interdisciplinary question)
	Multidisciplinarity
	Partial interdisciplinarity
	Conceptual interdisciplinarity
Transdisciplinarity	Transdisciplinarity
	Cross-disciplinarity
Conceptual interdisciplinarity	(True) interdisciplinarity
	Critical interdisciplinarity
	Full interdisciplinarity

transdisciplinarity (mode 2) can offer fundamentally different answers to questions of today's complex society. Mode 2 characterises the production of knowledge in an applied context, in which the interests of the societal, economic and political actors who constitute the research process are taken into account. They are involved from the beginning and contribute different types of competence and expertise in the research process (Gibbons et al. 1994).

A hierarchy should not be imposed on the terms disciplinarity, interdisciplinarity and transdisciplinarity. To do so would make little sense, for the organisation of a research process does not necessarily have to be transdisciplinary to be evaluated as 'good' or 'bad'. In other words, transdisciplinary research should not to be held in higher regard simply because it is transdisciplinary. Rather, the quality of the research depends on the extent to which the problem at hand is being dealt with in an appropriate manner. Disciplinarity and inter- or transdisciplinarity "are plausible valuations with respect to the operation of the research process in spite of their apparent contradiction, and both are crucially important. They are complementary rather than contradictory" (Weingart 2000: 29). Disciplinarity and inter- or transdisciplinarity are co-dependent and the knowledge as well as the quality of integration of knowledge and the broadening of perspectives are related to the distinctiveness of disciplinary boundaries.

Understanding Between Disciplines

After analysing the meaning of the two terms, the focus should be directed to the possibility of *understanding between disciplines*. This raises two major questions:

1. Questions about the *concept of disciplines* and the categorization of sciences, in particular questions of methodology and classification, the classification and delimitation of scientific disciplines. Disciplines create a framework of reference

and a system for scientific work. Kuhn (1970) has characterised this cognitive organisational structure using three elements: its underlying theory (generalization); idealized models and analogies (abstracted examples from real cases to ideal phenomena) and exemplars (specific instances of generalizations and models). Disciplines are social practices arising from human ideas and traditions. They form a communication network and are 'organised social groupings' (Whitley 1976). As Becher (1989) describes in his ironic interpretation of the world of science, disciplines or the 'tribes of academe' are characterised by "explicitly cultural elements: their traditions, customs and practices, transmitted knowledge, beliefs, morals and rules of conduct, as well as their linguistic and symbolic forms of communication and the meanings they share" (Becher 1989: 24).

2. Questions about *understanding between the disciplines* as well as across scientific boundaries. Language plays a key role here. "Differences in research methods, work styles, and epistemologies must be bridged in order to achieve mutual understanding of a problem and to arrive at a common solution. In transdisciplinary work, the language of stakeholders must also be recognized, although the language of target groups has not been viewed traditionally as a resource" (Thompson Klein 2004: 520).

Achieving understanding in an inter- or transdisciplinary team when fundamental terms are understood in radically different ways places high demands on the actors involved. "Interdisciplinarity conceived as communicative action rejects the naive faith that everything will work out if everyone just sits down and talks to each other. Decades of scuttled projects and programs belie the hope that status hierarchies and hidden agendas will not interfere, or the individual with the greatest clout or loudest voice will not dominate. The ideal speech situation assumes lack of coercion and equal access to dialog at all points" (Thompson Klein 2005: 44).

In the following the emphasis will be more on which factors have a decisive influence on the process of understanding. Inter- and transdisciplinary teams are a collective of humans and *group factors* have an important influence on the quality of their collaboration. The *flow of information* is a major element of every inter- and transdisciplinary act of communication. Communication can be described, as Niklas Luhmann puts it, as the "common actualisation of meaning" (Luhmann 1971: 42) and is essential for collaboration. Beyond the simple exchange of information, the goal and at the same time the highest quality level is a successful *integration of knowledge*. The integration of a diversity of relevant perspectives and understanding the complexity of a problem are the core challenges in inter- and transdisciplinary research and learning processes.

Group Factors

There are important factors at work in inter- and transdisciplinary groups, as in every other group. However each inter- or transdisciplinary team is not the same. Van Dusseldorp und Wigboldus (1994) have classified interdisciplinary groups according to the general type of discipline they are a member of. They describe teams made up of natural science researchers as 'narrow interdisciplinarity', as they make use of very similar paradigms and methods and share a common knowledge culture. They speak of 'broad interdisciplinarity' when teams are composed of natural scientists and social scientists and are also organised in different organisations. Such teams must be able to cope with different paradigms, methods and knowledge cultures.

In every type of group there are certain classic phenomena resulting from fears and uncertainties, for example the fear of a negative evaluation (Arrow et al. 2000). These aspects have a negative effect on performance and lead to individual group members not participating fully (Brodbeck and Frey 1999). In addition the following factors influence group processes and the exchange of knowledge:

- *The size of a team.* Previous research recommends that the number of group members be between 4 and 12, with between seven and nine as an optimal size for interdisciplinary work (Taylor 1975; Stankiewicz 1979). In general it can be said that groups with a constant number of members are best integrated. Groups that are too large over-complicate communication processes and there is a tendency to work at the level of the lowest commonly agreed upon denominator. Larger groups can still achieve good results when they are divided into sub-teams with an open communication structure. However this complicates the integration of knowledge as well as the delegation of responsibility (Thompson Klein 2005).
- *The degree of experience with collaboration* (Steinheider and Burger 2000). The more experience with (transdisciplinary) group work there is, the better the communication and collaboration in a team.
- *The status of group members* (Stasser et al. 1989). Status conflicts arise for a number of reasons, including gender, cultural background and race. A hierarchisation of disciplines leads to the creation of status differences, in which individual disciplines become dominant and see themselves as playing a leading role. Closely related to this phenomenon is the influence of *power*, which arises from differences in individual influence within a group and is a central factor in group processes.
- *The degree of familiarity* among individual group members. The development of meta-knowledge in the form of knowledge about the expertise of the others is influenced by knowledge of each other (Hollingshead 2000).
- *The leadership of the team.* The leader of an inter- or transdisciplinary team can be characterised in a number of different ways, from 'ringmaster' to 'boundary agent' to 'bridge scientist'. His tasks go beyond the translation of disciplinary perspectives and involve the integration of disciplinary perspectives (Anbar 1973). The management and communicative skills of leaders are a major variable of an inter- or transdisciplinary team's problem-solving ability (Thompson Klein 2005).

Effectiveness in groups characterised by a high degree of complexity, both regarding content and group composition, is very strongly dependent on the self-reflection skills within the group. This so-called 'group task reflexivity' (West 1996) is defined as "the extent to which group members overtly reflect upon the group's objectives, strategies and processes, and adapt them to current or anticipated endogenous or environmental circumstances" (West 1996: 559). Changes in the group are affected by the degree of group reflexivity and for group performance this kind of meta-reflection is a key factor. Especially for complex tasks characterised by uncertainty and ambiguity the degree of the group's self-reflexivity improves the results, that is group performance. In studies on the relationship between continuous learning in a group and team performance (Edmonson 1999), there is a significantly positive relationship between proactive critique of group collaboration and group performance (Gebert 2004: 25f.). Group performance in these cases involves a higher level of creativity, i.e. the development of creative ideas and innovations.

Along with these group-related factors influencing the success of collaboration it is necessary to study the *process of the exchange of knowledge.*

Flow of Information

Successful problem-solving in an inter- or transdisciplinary team depends on the willingness of group members to share their knowledge and other information in discussions (Larson et al. 1996). The more *non-shared* information, i.e. information (e.g. disciplinary knowledge) not possessed by all members that is introduced during the work, the more comprehensive the solution to the problem. There is a danger that known (shared) information rather than unknown (non-shared) information is contributed to the discussion (Stasser and Titus 1985). There are a number of explanations for this phenomenon, including that shared information has an advantage in being already accepted, is less conflict-laden and is more often repeated (Stasser et al. 1989). Already before the decision-making situation, group members tend to prefer certain types of subject matter and therefore evaluate received information that is contrary to these preferences as less relevant or credible (Greitemeyer and Schulz-Hardt 2003). Studies of problems relating to information exchange have revealed that so-called meta-knowledge about the expertise already within a group positively influences the exchange of non-shared information. If group members know who has which knowledge, i.e. who has expertise in which areas, the probability that non-shared information will be contributed to a discussion is increased, as is the quality of group decisions (Littlepage et al. 1997).

However it is not always the case that each group member develops a correct idea of the others' areas of expertise. It is also interesting that in the early phases of a discussion it is the shared information that dominates and the probability that non-shared information will be discussed increases with the length of the discussion (Larson et al. 1996). The time factor thus plays a considerable role in heterogeneous discussions. The members of inter- and transdisciplinary groups at first do not know what knowledge and expertise is represented in the group and possibly they also do not trust the new information, find it to be irrelevant or are unable to relate it to their own knowledge.

Social validation also has an influence on the expressive behaviour of group members. Information that has been negatively evaluated by the group flows into the decision-making process considerably less often than generally accepted information (Stewart and Stasser 1995). Paradoxically those who more often introduce shared information are considered to have more expertise than those who introduce non-shared information.

The so-called 'shared reality' approach (Levine et al. 2000) explains such phenomena by demonstrating how shared reality forms a reference point for the evaluation of information from other group members. Every type of group has "a common frame of reference. This common frame of reference is often described as the group's culture" (Levine and Moreland 1991: 258). Building on the common ground approach of Clark (1996), knowledge held in common is not only a condition for group action (Godemann 2008), but also represents 'social facts', i.e. what is judged by the group to be right or wrong. Shared reality thus determines the group's understanding of itself and influences decision-making processes. In the course of problem-solving processes the group develops its own shared reality of methods and strategies. Following Asch (1987) if a shared reality is to develop then it is necessary for the group members to perceive the actions of other members, interpret them and relate them to one's own activities. Levine and Higgins consider 'shared reality' as "the major contributor to group activity" (Levine and Higgins 2001: 34).

Shared reality and common ground essentially lead to shared mental models. The shared mental models of a group comprise the knowledge relating to group goals, characteristics, interaction patterns as well as role and behaviour patterns. In a word, they represent the shared knowledge necessary for collective action to take place. Mental models relate to a meta-knowledge that goes beyond the perspective of individual group members. In groups there may be individuals with greatly different perspectives on certain problem areas. Perspectives include opinions, attitudes, values and especially a cognitive structure that is related to the varying experiences and amounts of knowledge possessed by individuals. The development of group-related mental models presupposes the ability of group members to take on other perspectives.

Perspective taking can be understood as a process of understanding a person as part of a specific background. From a psychological point of view the taking of another perspective requires two mental processes. First there must be the concept of an outside perspective, i.e. there must be the realisation that another person has a different perspective. And second a process of thinking must take place that simulates and anticipates the perspective of the other. Only when both conditions are fulfilled can we speak of perspective taking (Flavell 1985). In order to accept and understand other perspectives it is necessary to undergo a reflection of one's own perspective. In inter- and transdisciplinary collaboration, truths from single disciplines lose their certainty or are relativised. 'Crises' are created on purpose and one's own discipline is questioned by outside perspectives. Ideally those involved gain a certain distance to what is considered established and are able to see things from another standpoint.

It remains to be seen how well experts, considering the availability of their own proven perspectives, are able to take into account the perspectives of their interaction partners. In a further step experts must be able to anticipate the lay perspective and then communicate their knowledge in a suitable fashion to laypeople. Contributions to the development of this kind of skill can be generated by inter- and transdisciplinary studies on sustainability (Godemann 2006).

In such learning processes competences can be enhanced that enable mutual learning. Those include learning to

- differentiate, i.e. learn different disciplinary perspectives;
- compare, i.e. compare knowledge of a different provenance and broaden one's own horizon;
- tolerate ambiguity, i.e. accept that there are different perspectives and solutions;
- synthesize and integrate, i.e. find compromises and develop solutions that are acceptable to all parties and are based on common ground;
- be sensitive, i.e. develop an awareness for ethical issues and the ability to promote sustainability.

The following section shows the steps in a process that would develop the ability to collaborate in inter- and transdisciplinary teams.

Enabling Integration

Thompson Klein (1990: 188) und Newell (2001: 248ff.) have formulated a framework for promoting knowledge integration, which will be drawn on and adapted in this section. It offers an orientation for collaboration in inter- and transdisciplinary groups and supports the process of knowledge integration. In order to develop teaching programmes, it would be appropriate to relate the individual steps to the methodology developed by the German Advisory Council on Global Change (WBGU 1996) for future-oriented research activities so that a basis for both content and methodology could be created. The first steps enable a transdisciplinary perspective of the problem and include:

- *Defining the problem*: Inter- and transdisciplinary problem-solving becomes necessary for the complex problems that research projects entail. The WBGU 'syndrome concept' entails a variety of central issues demonstrating the interdependence of global problems. Expert knowledge is used to identify global 'clinical patterns' that reflect critical changes (e.g. the global greenhouse effect, soil erosion, mass tourism). The approach specifies trends that are relevant to global change. These trends in human behaviour as it impacts on the environment form patterns of unsustainable development. Because of the linkages between disciplines, this approach relates the different areas of knowledge of the people involved (e.g. economics, political science, sociology, psychology, law, philosophy, engineering) in an inter- or transdisciplinary team.

- *Determining all of the relevant disciplines*: In order to capture the breadth of the problem and its varied aspects, it is critical to identify all of the schools of thought and societal actors that could contribute to solving the problem, or who are affected by the problem.
- *Developing a framework and appropriate questions to be investigated*: This step involves deciding what knowledge should be generated and how. Decisions must be taken as to which methods and theories will be used. The WBGU methodology offers a systematic approach to the analysis of non-sustainable trends in development. It enables complex interrelationships to be graphically portrayed and provides room for all relevant disciplines and actors to contribute their specific knowledge, whether of theories or methods.
- *Gathering current disciplinary knowledge*: In this process the goal is to search for new information, study the problem from the perspective of each discipline and generate disciplinary insights into the problem. "Difference, tension, and conflict are not barriers that must be eliminated. They are part of the character of interdisciplinary knowledge negotiation" (Thompson Klein 2005: 45).

The second step is the integration of knowledge through the construction of a more comprehensive perspective:

- *Creating common ground*: This is accomplished by looking for different terms with common meanings, or the same terms with different meanings. A discussion is initiated about disciplinary assumptions, leading to the creation of a common basis of knowledge as well as a common framework.
- *Constructing a new understanding of the problem*: The knowledge gained in the previous step can result in a comprehensive view of the problem and a broadening of perspectives. The WBGU approach permits understanding problems as systems and identifying the interrelationships within this system. In the process of integrating knowledge, it is the step *producing a model* (metaphor or theme) that captures the new understanding of inter- and transdisciplinary work.
- The complex description of a system can then be used as a starting point for finding ways out of non-sustainable trends to sustainable development. The finding of sustainable ways out of the problem is then to be understood as *testing the understanding by attempting to solve the problem*.

This type of knowledge exchange and knowledge integration provides inter- and transdisciplinary groups with a communication culture as well as a common cognitive frame of reference that permits not only the understanding of central concepts and terms but also cooperative action. The main challenge of knowledge integration is whether the different disciplines are able to cooperate to the extent that they provide different lenses for viewing the same phenomena instead of looking at different phenomena separately and then compiling the results. Successful communication depends on having a shared action context. Conversely this means that we can only interpret something foreign when we can draw on common forms or facts. This is also the background for the well-known remark by Wittgenstein that "If a lion could talk, we could not understand him" (PI II 223 in Glock 1996: 128). In a fashion then inter- and transdisciplinarity is a form of interculturality. Similar to ethnology, which had to first learn that other cultures exist and are not merely a preliminary form or a mixture of forms of European culture, so too must scientists learn that there are other disciplinary cultures and that they can provide an alternative perspective. "Changing one's perspective is like entering another culture" (Frank et al. 1992: 235).

References

Aboelela, S. W., Larson, E., Bakken, S., Carrasquillo, O., Formicola, A., Glied, S. A., Haas, J., & Gebbie, K. M. (2007). Defining interdisciplinary research: Conclusions from a critical review of the literature. *Health Service Research, 42*(1), 329–346.
Anbar, M. (1973). The bridge scientist and his role. *Research Development, 24*, 30–34.
Aram, J. D. (2004). Concepts of interdisciplinarity: Configurations of knowledge and action. *Human Relations, 57*(4), 379–412.
Arrow, H., McGrath, J. E., & Berdahl, J. L. (2000). *Small groups as complex systems: Formation, coordination, development and adaptation*. Thousand Oaks: Sage.
Asch, S. E. (1987). *Social psychology*. Oxford: Oxford University Press.
Becher, T. (1989). *Academic tribes and territories: Intellectual enquiry and the cultures of disciplines*. Milton Keynes: Open University Press.
Brodbeck, F. C., & Frey, D. (1999). Gruppenprozesse. In C. G. Hoyos & D. Frey (Eds.), *Arbeits- und Organisationspsychologie. Ein Lehrbuch* (pp. 358–372). Weinheim: Beltz.
Clark, H. H. (1996). *Using language*. Cambridge: Cambridge University Press.
Derry, S., Schunn, C., & Gernsbacher, M. (2005). *Interdisciplinary collaboration: An emerging cognitive science*. Mahwah: Lawrence Erlbaum Associates Inc.
Edmonson, A. (1999). Psychological safety and learning behavior in work teams. *Administrative Science Quarterly, 44*, 350–383.
Flavell, J. H. (1985). *Cognitive development* (2nd ed.). Englewood Cliffs: Prentice-Hall.
Frank, A., Schülert, J., & Nicolas, H. (1992). Interdisciplinary learning as social learning and general education. *European Journal of Education, 27*(3), 223.
Gebert, D. (2004). *Innovation durch Teamarbeit. Eine kritische Bestandsaufnahme*. Stuttgart: Kohlhammer.
Gibbons, M., Limoges, C., Nowotny, H., Schwartzmann, S., Scott, P., & Trow, M. (1994). *The new production of knowledge: The dynamics of science and research in contemporary societies*. London: Sage.
Glock, H. J. (1996). *A Wittgenstein dictionary* (The Blackwell philosopher dictionaries). Oxford, UK: Blackwell Reference.
Godemann, J. (2006). Promotion of interdisciplinary competence as a challenge for higher education. *Journal of Social Science Education, 5*(2), 51–61. Retrieved July 30, 2010, from www.jsse.org/2006-2/pdf/godemann_promotion.pdf.
Godemann, J. (2008). Knowledge integration: A key challenge for transdisciplinary cooperation. *Environmental Education Research, 14*(6), 625–641.
Greitemeyer, T., & Schulz-Hardt, S. (2003). Preference-consistent evaluation of information in the hidden profile paradigm: Beyond group-level explanations for the dominance of shared information in group decisions. *Journal of Personality and Social Psychology, 84*(2), 322–339.
Hirsch Hadorn, G., Hoffmann-Riem, H., Biber-Klemm, S., Grossenbacher-Mansuy, W., Joye, D., Pohl, C., Wiesmann, U., & Zemp, E. (Eds.). (2008). *Handbook of transdisciplinary research*. Dordrecht: Springer.
Hollingshead, A. B. (2000). Perceptions of expertise and transactive memory in work relationships. *Group Processes & Intergroup Relations, 3*(3), 257–267.
Jantsch, E. (1972). Towards interdisciplinarity and transdisciplinarity in education and innovation. In L. Apostel (Ed.), *Interdisciplinarity: Problems of teaching and research in universities* (pp. 97–121). Paris: OECD Organization for Economic Cooperation and Development.

Kockelmans, J. J. (1979). Why interdisciplinarity? In J. J. Kockelmans (Ed.), *Interdisciplinarity and higher education* (pp. 125–160). London: Pennsylvania State University.
Kuhn, T. (1970). *The structure of scientific revolutions* (2nd enlarged ed.) Chicago/London: University Press.
Larson, J. R., Christensen, C., Abbott, A. S., & Franz, T. M. (1996). Diagnosing groups: Charting the flow of information in medical decision-making teams. *Journal of Personality and Social Psychology, 71*(2), 315–330.
Lattuca, L. (2001). *Creating interdisciplinarity: Interdisciplinary research and teaching among college and university faculty*. Nashville: Vanderbilt University Press.
Lattuca, L. R. (2002). Learning interdisciplinarity: Sociocultural perspectives on academic work. *The Journal of Higher Education, 73*(6), 711–739.
Levine, J. M., & Higgins, E. T. (2001). Shared reality and social influence in groups and organizations. In F. Butera & G. Mugny (Eds.), *Social influence in social reality: Promoting individual and social change* (pp. 33–52). Seattle/Göttingen: Hogrefe & Huber.
Levine, J. M., & Moreland, R. L. (1991). Culture and socialization in work groups. In L. B. Resnick & J. M. Levine (Eds.), *Perspectives on socially shared cognition* (pp. 257–281). Washington, DC: American Psychological Association.
Levine, J. M., Higgins, E. T., & Choi, H.-S. (2000). Development of strategic norms in groups. *Organizational Behavior and Human Decision Processes, 82*(1), 88–101.
Littlepage, G., Robison, W., & Reddington, K. (1997). Effects of task experience and group experience on group performance, member ability, and recognition of expertise. *Organizational Behavior and Human Decision Processes, 69*(2), 133–147.
Luhmann, N. (1971). Sinn als Grundbegriff der Soziologie. In J. Habermas & N. Luhmann (Eds.), *Theorie der Gesellschaft oder Sozialtechnologie – Was leistet die Systemforschung?* (pp. 25–100). Frankfurt am Main: Suhrkamp.
Mittelstraß, J. (2001). Learning live together: New challenges to education and research in a global economy. *Prospects, 31*(3), 393–398.
Mittelstraß, J. (2002). *Transdisciplinarity – New structures in science. Innovative structures in basic research* (pp. 43–54). Paper presented at the Ringberg-Symposium 2000, ed. Max-Planck-Gesellschaft, München.
Newell, W. (2001). A theory of interdisciplinary studies. *Issues in Integrative Studies, 19*, 1–25.
Nilles, J. M. (1975). Interdisciplinary research management in the university environment. *Journal of the Society of Research Administration, 6*(9), 9–16.
OECD. (1972). *Problems of teaching and research in universities*. Paris: OECD Organization for Economic Cooperation and Development.
OECD. (1998). *Interdisciplinarity in science and technology. T. Directorate for science, and industry*. Paris: OECD Organization for Economic Cooperation and Development.
Stankiewicz, R. (1979). The effects of leadership on relationship between the size of research groups and their performance. *R&D Management, 9*, 207–212.
Stasser, G., & Titus, W. (1985). Pooling of unshared information in group decision making: Biased information sampling during discussion. *Journal of Personality and Social Psychology, 48*(6), 1467–1478.
Stasser, G., Taylor, L. A., & Hanna, C. (1989). Information sampling in structured and unstructured discussions of three- and six-person groups. *Journal of Personality and Social Psychology, 57*(1), 67–78.
Steinheider, B., & Burger, E. (2000). Kooperation in interdisziplinären Teams. In Gesellschaft fuer Arbeitswissenschaft e.V (Ed.), *Komplexe Arbeitssysteme – Herausforderungen für Analyse und Gestaltung* (pp. 553–557). Dortmund: GfA Press.
Stewart, D. D., & Stasser, G. (1995). Expert role assignment and information sampling during collective recall and decision making. *Journal of Personality and Social Psychology, 69*(4), 619–628.
Taylor, J. B. (1975). Building an interdisciplinary team. In S. Arnstein & A. Christakis (Eds.), *Perspectives on technology assessment* (pp. 45–60). Jerusalem: Science and Technology Publisher.

Thompson Klein, J. (1990). *Interdisciplinarity: History, theory, and practice*. Detroit: Wayne State University Press.
Thompson Klein, J. (1996). *Crossing boundaries: Knowledge, disciplinarities, and interdisciplinarities*. Charlottesville/London: University Press of Virginia.
Thompson Klein, J. (2004). Prospects for transdisciplinarity. *Futures, 36*, 515–526.
Thompson Klein, J. (2005). Interdisciplinary teamwork: The dynamics of collaboration and integration. In S. Derry, C. D. Schunn, & M. A. Gernsbacher (Eds.), *Interdisciplinary collaboration: An emerging cognitive science* (pp. 23–50). Mahwah: Erlbaum.
Thompson Klein, J. (2010). *Creating interdisciplinary campus cultures: A model for strength and sustainability*. San Francisco: Wiley.
Thompson Klein, J., Frodeman, R., & Mitcham, C. (2010). *The oxford handbook of interdisciplinarity*. Oxford: Oxford University Press.
Van Dusseldorp, D., & Wigboldus, S. (1994). Interdisciplinary research for integrated rural development in developing countries: The role of social science. *Issues in integrative Studies, 12*, 93–138.
WBGU (German Advisory Council on Global Change) (1996). *World in Transition: The Research Challenge*. Annual Report 1996. Berlin: Springer.
Weingart, P. (2000). Interdisciplinartiy: The paradoxical discourse. In P. Weingart & N. Stehr (Eds.), *Practising interdisciplinarity* (pp. 25–41). Toronto: University Press.
Weingart, P., & Stehr, N. (Eds.). (2000). *Practising interdisciplinarity*. Toronto: University of Toronto Press.
West, M. A. (1996). Reflexivity and work group effectiveness: An conceptual integration. In M. A. West (Ed.), *Handbook of work group psychology* (pp. 555–579). Chichester: Wiley.
Whitley, R. (1976). Umbrella and polytheistic scientific discipline and their elites. *Social Studies of Science, 6*, 471–497.

Part II
Framework of Sustainability Communication

Chapter 5
Sociological Perspectives on Sustainability Communication

Karl-Werner Brand

Abstract From a sociological perspective, social communication has a key role in the stabilisation and change of institutional practices as well as in sustainability communication. As this promotes the development and dissemination of new institutional practices oriented towards a vision of sustainability, the analysis of the relationship between public communication and institutional change is of particular importance. This chapter attempts to answer four questions: What can be learned about this relationship from a number of sociological approaches? What special frames characterise sustainability discourse in Germany? What institutional practices are thus advantaged? And what role does the social embedding of everyday actions in lifestyle milieux have for the implementation of widely accepted environmental norms?

Keywords Environmental sociology • Institutional practices • Sustainability discourse in Germany • Lifestyle • Milieux

"Society is unthinkable without communication, but communication is also unthinkable without society" (Luhmann 1997: 13). For Luhmann, communication is the basic operation that produces and reproduces societies. Ecological and sustainability problems also only exist as a social problem to the extent that there is communication about it. If communication is given such a constitutive role for the development and identity of society, this does not automatically mean that we share the premises of Luhmann's systems theory, which follows Maturana's autopoiesis model. Societies and social sub-systems need not be understood as 'self-referentially

K.-W. Brand (✉)
Technical University of Munich, Munich, Germany
e-mail: karl-werner.brand@tum.de

closed' communication systems even if we share the assessment that economic, political or social movement actors primarily perceive and analyse environmental and sustainability problems according to their own internal 'rationalities'. Sociology has a variety of different theoretical approaches to the understanding of social processes, including systems theory, action theory, symbolic interactionism, neo-Marxism or discourse theory, all of which can be used to study sustainability communication from different angles.

Communication and Institutional Practices: Sociological Approaches

Although competing with each other, most approaches are no longer seen as exclusive schools today. In problem-oriented research especially there is a trend towards productive eclecticism. This also goes for theoretical debates where synthesising approaches have met with growing resonance since the 1980s. One of the best known is Giddens' theory of structuration (1984), which attempts to overcome the gap between action and structure through a perspective that emphasizes their reciprocal reproduction. What Giddens calls 'duality of structure' refers to the fact that institutional structures not only constrain but also enable social action. If the function of institutions is broadly understood as regulating social life, providing interpretations of reality that give meaning and identity, offering adequate strategies for problem solution, governing the division of power and resources, norming patterns of behaviour and sanctioning deviance, then institutions both create the preconditions of ordered social life and limit the scope for possible modes of social action. Institutions are, however, able to structure social life only to the extent that social actors reproduce them in everyday practices, thus confirming and re-confirming their validity.

If this reciprocal process of constituting action and structure, everyday practices and systemic processes, is combined with the insight that communication is the basic medium for constructing social reality, then Giddens' approach can also be given a symbolic-interactionist twist. This research perspective assumes that humans are able to act because of the meanings they attribute to situations, institutions, things, nature etc., whereby these meanings are continually adapted to a particular field of action (Blumer 1969; Jonas 1987). Meanings are not only negotiated interpersonally and situationally, they also achieve a normative, 'objective' power within the process of institutionalisation (Berger and Luckmann 1966). They structure our – mostly latent – everyday knowledge and deliver categorisations through which we try to find our bearings in reality and attempt to influence it. Finally, they also give us a basis for legitimising, or criticising, existing institutions.

Symbolic interactionism and Giddens' theory of structuration converge in important aspects. Dominant interpretations of reality and institutionalised social practices rely on continual communicative reproduction in everyday life in order to exercise their orienting and normative functions. This includes a continual symbolisation and ritual staging of the implicit *idée directrice* (M. Hauriou), the ideas governing a given institution. If the validity and stability of institutions depend to a large extent on how these 'governing ideas' resonate with the beliefs and practices of everyday life, then the possibility of institutional change depends on a loss of their practical plausibility in the face of changing life contexts and new, problematic situations. Institutional change does not, however, happen on its own. Institutions are closely connected with power relations. It is only the active questioning of the governing ideas of institutions in public debate and the successful mobilisation of competing interpretive frames, myths and symbols that can withdraw their legitimation. Whether and to what extent this is successful is a question of the discursive power of the competing collective actors. This also holds true for a transformation toward sustainability (Alexander 2009; Dingler 2003; Feindt and Oels 2005; Hajer 1995; Hajer and Versteeg 2005).

It is therefore interesting to examine sustainability communication from the perspective of discourse theory. There are roughly two versions of this approach (Jørgensen and Philipps 2002; Keller et al. 2001; Keller 2004; Phillips and Hardy 2002), a post-structuralist and a symbolic-interactionist or phenomenological version, in both of which the interrelationship of discourse and institutional practices plays a central role.

Post-structuralist approaches (Fairclough 2003; Howarth 2000; Laclau and Mouffe 1985) examine – often with reference to Foucault – the rule-bound structures of knowledge that discourses are based upon. In a Foucauldian sense, discourses are constitutive of reality, not only in a symbolic but also in a practical, material way. This productive, reality-constituting effect of discourses is the result of the power present in all forms of social interaction. "The Foucauldian understanding of discourse implies a conception of power as constitutive and productive. (…) Power is understood as a web of force relations made up of local centres of power around which specific discourses, strategies of power and techniques for the appropriation of knowledge cluster" (Feindt and Oels 2005: 164). Post-structuralist approaches thus move to the foreground the systematic interrelationship of power and the production of knowledge, as well as the disciplining aspect of discourses. Discourses define the kind of questions that may be posed; they determine the group of individuals that are authorised to take part in certain discourses; they contain ways to discipline; and they determine the conditions under which certain discourses can take place.

Symbolic-interactionist or *phenomenological* approaches (Gamson 1988; Hajer 1995; Gusfield 1981; Keller 2005), on the other hand, tend to highlight the interactive dynamics of the communicative construction of reality. In this perspective, conflict discourses are understood as controversially structured fields of symbolic interaction in which a variety of actors struggle to establish their respective interpretation of problems, their causes and remedies. These discursive struggles usually are structured by competing 'frames' (Gamson 1988) or organised around two competing 'storylines' which create order in the confusing array of arguments and allow heterogeneous positions to rally into clear-cut 'discourse coalitions' (Hajer 1995).

In modern societies, the symbolic struggles for cultural hegemony are carried out primarily in the arena of mass media. Taking the ecological debate as an example, a number of studies have shown how the specific selectivity of media reporting influences the dynamics of public discourse (Alexander 2009; Brand et al. 1997; Cox 2006; Hansen 1993; Neuzil and Kovarik 1996). The public media debate, however, is not the only level on which adversaries communicate with each other. Many conflict discourses take place initially, or largely, in the restricted domain of a specialist audience. This is especially true for debates about how to specify sustainable development in the diverse fields of action, such as mobility, agriculture, housing etc.

Sustainability Communication as a Controversially Structured Field of Discourse

What insights for sustainability communication can be gained from these different sociological approaches to the analysis of the relationship between discourse and institutional practices?

- A basic insight is that public communication is of central importance for establishing new institutional practices that are oriented toward the guiding idea of sustainability. The interpretations that become dominant in public discourses not only let certain institutional forms of regulation seem appropriate, they also allow the interests and power structures connected with them to appear legitimate – while others are rendered inappropriate and illegitimate.
- The approaches outlined above are also in agreement in that institutional change towards sustainability requires resonant problem framings that are able to mobilise relevant parts of the public so that the governing ideas and storylines of existing institutional practices can be called into question. It is a critical weakness of sustainability communication that this has only been achieved to a very limited extent: the traditional discourse of economic growth remains dominant. This is largely due to the fact that although the concept of sustainability meets with broad general approval, its diffuseness and the various possibilities of interpreting it deprive it of the ability to mobilise a broader, integrative reform movement.
- A third insight relates to the fact that sustainability communication can best be understood as a discursive field in which competing actors struggle for the power to frame sustainability problems in a publicly accepted way. To be sure, this discourse field is integrated by a diffuse norm of global and intergenerational fairness. There is also a large measure of agreement that sustainability problems can only be solved by systematically linking ecological, economic and social aspects of development. Nevertheless, sustainability remains a controversial concept, behind which there are different interests, conflicting views of the world and of nature as well as diverse understandings of development and societal regulation (see Dingler 2003; Dobson 2000; Dryzek 1997; Jacobs 1999; McManus 1996; Sachs 1997). There are basic controversies on ecological, social and economic questions of sustainable development, but each issue also produces a

somewhat different constellation of conflicting parties with different opportunities to forge new cross-cutting discourse coalitions and political alliances.

The sociological perspective thus emphasizes the fact that the sustainability debate is not just about looking for the best solutions of sustainability problems but also about a comprehensive norm-building process, a restructuring of social interpretations of reality and institutional practices. If specific ways of framing problems define the range of possible and legitimate ways of solving them, then the question of which frames, images, and metaphors gain public acceptance is of vital importance for the kind of policies and measures adopted.

The German Discourse on Sustainable Development: An Example

This sociological perspective will be illustrated in the following section using the example of the German discourse on sustainable development. This outline refers to an empirical study of the German sustainability debate at its formative stage from the mid- to the late-1990s (Brand and Jochum 2000). The goal of this study was a qualitative reconstruction of the central frames of this debate through the use of key documents, i.e. a number of programmatic studies on sustainable development published in the mid-1990s and position papers from major political, economic and social actors, complemented by a media analysis.

At the beginning of the 1990s the environmental debate in Germany was in a crisis – at least as perceived by environmental actors. The late 1980s had seen a general greening of political debates in which the concept of ecological modernisation had advanced to being a generally accepted reference point for societal innovation. But ecological concerns lost their political importance to traditional growth and cost arguments in the wake of the social and economic problems following German reunification in 1990. There was a feeling of resignation among environmental groups and a search for new strategic approaches was launched (Brand 1999). It was in this phase of disorientation that the reception of the international debate on sustainable development took place, especially inspired by the Agenda 21 adopted by the Rio Conference in 1992. In the following years a great variety of concepts for implementing the idea of sustainable development was developed, not only as part of flourishing local Agenda 21 initiatives but also in many social, economic, political and scientific organisations. It should come as no surprise that there were different perspectives about how to take action, which was as true for Germany as for all other Western countries (cf. Baker et al. 1997; Lafferty and Meadowcroft 2000). A central position in the German debate was held by two reports from the Enquete Commission "Protection of Mankind and Environment" of the German Federal Parliament (1994, 1998) and a study on 'Sustainable Germany' (Sachs et al. 1998) by the Wuppertal Institute for Climate, Environment and Energy.

Which positions have emerged in the following years in the German debate on sustainability development? Although all actors acknowledge, or at least pay lip service to, the demand for a more sustainable way of life and the need for an integrative

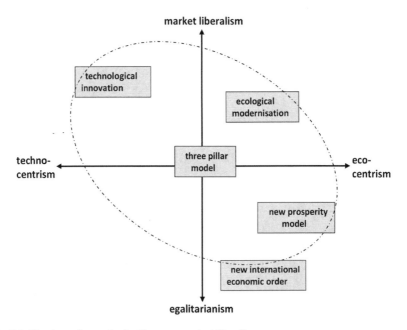

Fig. 5.1 Dominant frames in the German sustainability discourse

approach, there are major differences that are due not only to the conflicting interests of the actors involved but also to their diverging views of society and nature. These controversial positions can be located in a discourse field structured by two axes (Fig. 5.1).

The *vertical axis* distinguishes different understandings of society and justice, with 'market liberalism' and 'egalitarianism' at its two ends. Business representatives generally see the free development of a globalized economy and the liberalisation of world trade as a crucial condition for sustainable development while international solidarity movements take the opposite view: they regard the power structures and the dynamics of global capitalism as the central motor of non-sustainable development and call for a new, more just world economic order. The *horizontal* axis shows different models of the relationship between society and nature, with 'techno-centrist' and 'eco-centrist' at its two ends. While the eco-centrist side represents the position of 'respect for nature' and calls for a soft 'adaptation to natural cycles' instead of 'violent' technological interventions, those groups closer to the techno-centrist pole see technological innovations as the decisive precondition for sustainable development.

Until the change of government from the conservative-liberal government under Chancellor Helmut Kohl to the red-green government of Gerhard Schröder in 1998, the development of the German debate was dominated by positions that emphasized both the upper left-hand and the lower right-hand quadrant. The frame favoured by business and the Kohl government, 'sustainability through technological innovation', stood in opposition to 'sustainability through new models of prosperity', the interpretation formulated by the Wuppertal Institute and anchored in a spectrum of environmental movements and development organisations. The position typical of the upper right-hand quadrant – and exemplified by the frame represented by the Advisory Council on the Environment, 'sustainability through ecological modernisation' – at first found little resonance in the public debate but then, after the change of government, became part of the official policy of the Ministry for the Environment. Since the end of the 1990s, as a result of the spectacular, internationally coordinated protests by opponents of globalisation, the frame 'sustainable development through a new international economic order' has received greater public response too.

These opposing interpretations were mediated by the procedural, integrative sustainability concept of the Enquete Commission "Protection of Mankind and the Environment", which interpreted sustainable development as an open, participative trade-off process between ecological, social and economic dimensions ('three-pillar model'). This mediating frame provided an integrative foundation for practical cooperation and strategic alliances of diverse social groups for the advancement of sustainable development. This procedural, multi-dimensional interpretation of sustainability reached a dominant position in the German discourse on sustainable development towards the end of the 1990s.

The price for this procedural, integrative understanding of sustainability, however, is a loss of clarity. The term *sustainability* has tended to become a catchword, meaning anything and everything. It no longer provokes and polarises and is thus hardly present in the public media – in contrast to the debates among committed sustainability experts (Brand 2000). The focus on the integrative trade-offs between different interests and perspectives also largely hides the conflict and power dimension of sustainable development. Windows of opportunity for fundamental changes thus only open by chance, through more or less dramatic events. For example, the first case of BSE was heatedly discussed in Germany in November 2000 and opened up the opportunity for a radical change in German agricultural and consumer policy, which brought about considerable dynamism in the organic food market (Brand 2006). A new window of opportunity for a more radical shift in German climate policies opened in spring 2007 in response to the fourth IPCC report on climate change, which found great resonance in the mass media in Germany. The climate issue, however, disappeared from the political agenda very quickly when the economic consequences of the global financial crisis became a top issue in the following year. The dependence on catastrophes, scandals and dramatic media events thus cannot provide a reasonable basis for a 'strategic', long-term sustainability policy (cf. Jänicke and Jörgens 2000).

In the framework of this integrative, procedural concept of sustainability, some other closely interrelated structural issues remain in the background: economy, work and gender, all three of which are basic elements of the industrial growth model. Whenever there is talk of an 'ecological modernisation' of the economy, the structural framework that is implicitly assumed is not only that of a growth-oriented capitalist economy but also that of a *formal* market economy. The sphere of reproductive economics remains hidden, as it is traditionally the province of women and considered second rate, even though the reproduction, the 'sustainability' of social life, is largely dependent on the smooth functioning of this sphere (Biesecker and Hofmeister 2006).

A central shortcoming of the sustainability debate is connected to this very aspect: work is schematised exclusively in the form of gainful employment. In the context of the debate about 'new models of prosperity', research has been carried out into a new understanding of work involving a balance between gainful employment, family care, self-employment and community work (Brandl and Hildebrandt 2002; Spangenberg 2003; Stahmer and Schaffer 2006). This new conception of work-life balance, however, has so far not been able to call into doubt the hegemonic debate fixated on the traditional triad of growth, gainful employment and consumption.

The structure of the German sustainability discourse thus favours and legitimates a certain pattern of institutional practices dealing with sustainability problems. Fading out relevant dimensions of these problems from the public debate leads to corresponding gaps in the options for practical action.

Sustainability Discourse, Lifestyle and Everyday Practices

The result of both the low presence in the mass media as well as the vagueness of the term 'sustainable development' is that it means little to a broader public audience.[1] Thus in Germany, energy-saving, environmentally-friendly transportation or organic food campaigns are typically framed without reference to 'sustainability'. Terms as 'energy-saving', 'organic', 'nature', 'fairness', 'health', 'countryside', 'region' etc. evoke symbolic associations that have considerably more mobilisation potential. The question then is whether the sustainability discourse is able to influence everyday life at all.

In fact, on a general level, it is not the integrative, multi-dimensional concept of sustainable development but its more specific understanding as 'ecological sustainability' that has had some influence on public debates in Germany. It stimulated a reframing of the public perception of environmental problems as complex interrelated global problems and introduced a sense of long-term, intergenerational responsibility into the debate. With the general shift of the environmental debate in the 1990s towards 'sustainable consumption', the social aspects of responsible consumption (fair trade, child labour etc.) became more important in Germany as well. However, which of these aspects of 'sustainable' behaviour is taken up in which field of action (food, transportation, waste separation, energy saving etc.) depends in Germany, as in all other countries, on specific national as well as individual factors.

---

[1] To the extent that the term 'sustainable development' is known at all in Germany (roughly 20%), it is associated with ideas of 'ecological economizing' or 'responsibility for future generations' (Kuckartz and Rheingans-Heintze 2006: 16f.).

On the one hand, the selectivity of public attention to certain issues is influenced both by the way a given country is affected by particular problems and by its specific cultural traditions, which create a very different responsiveness for (different kinds of) environmental, social and technological problems and risks (e.g. Goodbody 2002; Rootes 2007; Selin 2003). On the other hand, in all countries the link between environmental awareness and behaviour is generally not only rather weak, it also shows a very inconsistent pattern, even among people with strong pro-environmental beliefs (e.g. de Haan and Kuckartz 1996; Diekmann and Preisendörfer 1998; Kollmuss and Agyeman 2002; Middlemiss and Young 2008). These findings are not very surprising considering the difficulty of making actual choices in line with strict ecological criteria in a society geared toward economic growth and material affluence. Institutional efforts to establish new and more environmentally friendly practices encounter a host of structural barriers. Individual choices are moreover complicated by incomplete and overly complex information, adverse price incentives, poor supply, insufficient infrastructural arrangements, practical inconveniences, contradictory interests, values and norms, which all render the ideal of consistent pro-environmental behaviour a particularly intricate venture.

This does not militate against a broad dissemination of ecological norms in most Western countries. These norms can also be strengthened in individual areas of behaviour through lifestyle trends, as can be seen from the great media resonance that 'LOHAS' (Lifestyle of Health and Sustainability) has met with since 2007. This trend influences above all the market (Wenzel et al. 2007). LOHAS are, however, no less inconsistent in their behaviour. Indeed, it would be an illusion to assume that this specific mixture of wellness and sustainability could be equally attractive to all social milieux. On the contrary, what is typical of all social groups is that expectations of ecological behaviour are integrated into everyday life only very selectively. The answer to why this is so varies according to the theoretical approach taken.

In the case of psychology, such decisions are attributed to attitudes, values, norms and motivations (van Kasteren 2008), in economic or in rational choice theory to 'rational' cost-benefit calculations, which typically entrap individual environmental behaviour in 'social dilemma' situations (Diekmann 1996). Sociological studies criticize the individualistic assumptions of these approaches. In contrast, they emphasize the social and cultural embeddedness of environmental behaviour and focus on the symbolic meaning of lifestyles and consumption (Bourdieu 1984; Featherstone 1991; Reusswig 1994).

The lifestyle concept is used in a number of different ways in sociology with a central controversy centring on the question of how closely lifestyles are linked with socio-economic life situations or class structures. Are lifestyles, as Bourdieu (1984) argues, the cultural practice of a 'class habitus', which itself is determined by the relational position of actors in the hierarchically structured fields of economic, social and cultural capital? Or have accelerated processes of social disembedding and individualisation produced a more reflexive pattern of life that demands a more active shaping and enactment of lifestyles, as is claimed by Beck, Giddens and other theorists of post-modernity or 'reflexive modernity' (Beck 1992; Giddens 1991)?

However, even in the latter case there is no presumption of a random variation of individual lifestyles. Rather, it is possible in all countries to identify social groups with similar ideas about life and ways of living which can be sorted into different life-style milieux and can be positioned in the 'social space' (Bourdieu) of a given society.

The members of these milieux share basic value orientations, have similar preferences in taste and styles of consumption, similar attitudes towards work, family and leisure but also towards the environment and politics. These milieu-specific commonalities in sports, leisure or cultural areas reflexively strengthen social identities and serve at the same time to mark social distinction (Schwenk 1996). Depending on the dynamics of social change, these milieux show a greater or lesser degree of stability, may change or are created anew. In the 1980s and 1990s a number of empirical studies in Germany developed such milieu typologies (e.g. Flaig et al. 1993; Schulze 1992; Vester et al. 1993, 1995) that identified a considerable shift in the forms of social inequality as well as a greater differentiation in new lifestyles. Some of these typologies, especially the 'SINUS Milieus' (Sinus Sociovision), are used in a number of very different contexts, from market research to political attitude surveys and environmental research in the social sciences. Representative surveys conducted at regular intervals by the German Federal Environmental Agency on 'Environmental awareness and behaviour in Germany', for example, investigated the specific environmental attitudes and behaviour in ten SINUS milieux in Germany in 2008 (Umweltbundesamt 2009).

Such lifestyles are a kind of filter for the translation of sustainability discourses into the everyday life of different social milieux (Rink 2002). They determine which aspects of this debate – together with which implicit conditions for action – find a high or low resonance. In the extreme case of 'ecological pioneers', ecological norms can also become the central, organising principle of their lifestyle. The cognitive side of the selective, group-specific internalisation of ecological norms in everyday consciousness can be reconstructed in the form of typical 'environmental mentalities' (for Germany, see Poferl et al. 1997; Brand et al. 2003). As to the behavioural aspect of lifestyles, empirical research has focused on the question of which basic action motives provide the closest link of individual lifestyles to sustainable consumption (for example, ECOLOG 1999; Kleinhückelkotten 2005) and how mobilisation campaigns or political incentive systems, for instance, for sustainable mobility, living or nutrition can make use of these insights (Götz 2007; Empacher and Hayn 2005; Schultz and Stieß 2008).

The expectations placed on such target-group specific dissemination strategies are nevertheless mostly too high. They must be tailored very selectively in order to reach their specific target group, which is a very resource-consuming exercise and can usually only be done as part of commercial product marketing. In addition, it is often overlooked that for the change of consumption patterns, inconspicuous 'ordinary consumption' is more important than the 'conspicuous' aspects of consumption that play the dominant role in the distinction of lifestyles (Gronow and Warde 2001; Shove and Warde 2002). To be sure, these conspicuous aspects have considerable ecological implications too. What has a much greater effect on the overall sustainability of social life are, however, the given sociotechnical 'systems of provision' (energy and water supply, settlement structures, mobility systems, construction standards etc.), existing market structures as well as cultural expectations and standards of 'normality' (Shove 2003; Southerton et al. 2004).

Summary

This limitation of a lifestyle-related communication approach to the dissemination of sustainable consumption does not question the importance of public sustainability discourses. Without the presence of a controversial discourse on global environmental problems and non-sustainable development paths in the mass media, there would be no pressure on either institutional or private actors to take action. Even if this discussion is not led under the heading of 'sustainable development' in the broad public, the public framing of the constitutive problems of this debate limits the scope in which practical changes can take place. There is another fundamental insight of modern sociology that can be utilised for an understanding of the problems of sustainability transition: the insight that the diverse spheres or sub-systems of social life follow their own internal rationalities. It is not only the broad spectrum of conflicting interests and diverging worldviews but also these different social rationalities that account for the translation of the general, widely accepted idea of 'sustainable development' into very specific and often contradictory action programmes (e.g. Luhmann 1989). This selective and contradictory 'translation process' happens in a similar way at the level of everyday life. Here it is the variety of lifestyles that translates the general postulate of an ecologically and socially responsible behaviour into very selective, milieu-specific patterns of problem awareness and consumption. It is this socio-cultural selectivity that gives the public controversies on sustainability issues a specific resonance in the life world of people. Both, the competing frames of sustainability problems in public discourses and their everyday cultural resonance determine the chances of a more neoliberal or egalitarian, a more techno- or eco-centrist strategy of sustainable development.

References

Alexander, R. (2009). *Framing discourse on the environment: A critical discourse approach.* London: Routledge.
Baker, S., Kousis, M., Richardson, D., & Young, S. (Eds.). (1997). *The politics of sustainable development: Theory, policy and practice within the European Union.* London: Routledge.
Beck, U. (1992). *Risk society: Towards a new modernity.* London: Sage.
Berger, P. L., & Luckmann, T. (1966). *The social construction of reality: A treatise in the sociology of knowledge.* Garden City, NY: Anchor Books.
Biesecker, A., & Hofmeister, S. (2006). *Die Neuerfindung des Ökonomischen: Ein (re)produktionstheoretischer Beitrag zur sozial-ökologischen Forschung.* München: Oekom.
Blumer, H. (1969). *Symbolic interactionism: Perspective and method.* Berkeley: University of California Press.

Bourdieu, P. (1984). *Distinction: A social critique of the judgement of taste*. Cambridge, MA: Harvard University Press.
Brand, K.-W. (1999). Dialectics of institutionalisation: The transformation of the environmental movement in Germany. In C. Roots (Ed.), *Environmental movements: Local, national and global* (pp. 35–58). London: Frank Cass.
Brand, K.-W. (2000). Kommunikation über nachhaltige Entwicklung, oder: Warum sich das Leitbild der Nachhaltigkeit so schlecht popularisieren lässt. *Sowi-onlinejournal, 1*(1), 1–18.
Brand, K.-W. (2006). *Die neues Dynamik des Bio-Markts. Folgen der Agrarwende im Bereich Landwirtschaft, Verarbeitung, Handel, Konsum und Ernährungskommunikation*. München: Oekom.
Brand, K.-W., & Jochum, G. (2000). *Die Struktur des deutschen Diskurses zu nachhaltiger Entwicklung*. München: MPS-Texte 1/2000 [From http://www.sozialforschung.org/wordpress/wp-content/uploads/2009/09/kw_brand_deutscher_nachh_diskurs.pdf].
Brand, K.-W., Eder, K., & Poferl, A. (1997). *Ökologische Kommunikation in Deutschland*. Opladen: Westdeutscher.
Brand, K. -W., Fischer, C. & Hofmann, M. (2003). *Lebensstile, Umweltmentalitäten und Umweltverhalten in Ostdeutschland*. UFZ-Texte 11/2003, Leipzig-Halle.
Brandl, S., & Hildebrandt, E. (2002). *Zukunft der Arbeit und soziale Nachhaltigkeit*. Opladen: Leske+Budrich.
Cox, R. (2006). *Environmental communication and the public sphere*. London: Sage.
De Haan, G., & Kuckartz, U. (1996). *Umweltbewusstsein: Denken und Handeln in Umweltkrisen*. Opladen: Westdeutscher.
Diekmann, A. (1996). Homo ÖKOnomicus. Anwendungen und Probleme der Theorie rationalen Handelns im Umweltbereich. In A. Diekmann & C. C. Jäger (Eds.), *Umweltsoziologie* (pp. 89–118). *Kölner Zeitschrift für Soziologie und Sozialpsychologie*, Sonderheft 36. Opladen: Westdeutscher.
Diekmann, A., & Preisendörfer, P. (1998). Environmental behavior. Discrepancies between aspirations and reality. *Rationality and society, 10*(1), 79–102.
Dingler, J. (2003). *Postmoderne und Nachhaltigkeit. Eine diskurstheoretische Analyse der sozialen Konstruktion von nachhaltiger Entwicklung*. München: Oekom.
Dobson, A. (2000). *Green political thought* (3rd ed.). London: Routledge.
Dryzek, S. (1997). *The politics of the earth: Environmental discourses*. Oxford: Oxford University Press.
ECOLOG-Institut. (1999). *Wegweiser durch Soziale Milieus und Lebensstile für Umweltbildung und Umweltberatung*. Hannover: ECOLOG-Institut.
Empacher, C., & Hayn, D. (2005). Ernährungsstile und Nachhaltigkeit im Alltag. In K.-M. Brunner & G. Schönberger (Eds.), *Nachhaltige Ernährung*. Frankfurt/New York: Campus.
Enquete Commission of the German Bundestag "Protection of Mankind and Environment". (1994). *Shaping industrial society*. Bonn: Economica.
Enquete Commission of the German Bundestag "Protection of Mankind and Environment". (1998). *Concept sustainability: From the model to the implementation. Final report*. Bonn: Economica.
Fairclough, N. (2003). *Analysing discourse: Textual analysis for social research*. London: Routledge.
Featherstone, M. (1991). *Consumer culture and postmodernism*. London: Sage.
Feindt, P. H., & Oels, A. (2005). Does discourse matter? Discourse analysis in environmental policy making. *Journal of Environmental Policy & Planning, 7*, 161–173.
Flaig, B., Meyer, T., & Ueltzhöffer, J. (1993). *Alltagsästhetik und politische Kultur. Zur ästhetischen Dimension politischer Bildung und politischer Kommunikation*. Bonn: Dietz.
Gamson, W. A. (1988). Political discourse and collective action. *International Social Movement Research, 1*, 219–46.
Giddens, A. (1984). *The constitution of society: Outline of the theory of structuration*. Cambridge: Polity Press.
Giddens, A. (1991). *Modernity and self-identity: Self and society in late modern age*. Cambridge: Polity Press.

Götz, K. (2007). Mobilitätsstile. In S. Oliver et al. (Eds.), *Handbuch Verkehrspolitik* (pp. 760–784). Wiesbaden: VS Verlag für Sozialwissenschaften.
Goodbody, A. (2002). *The culture of German environmentalism. Anxieties, visions, realities.* New York and Oxford: Berghahn.
Gronow, J., & Warde, A. (Eds.). (2001). *Ordinary consumption.* London: Routledge.
Gusfield, J. (1981). *The culture of public problems.* Chicago: The University of Chicago Press.
Hajer, M. (1995). *The politics of environmental discourse: Ecological modernization of the political process.* Oxford: Clarendon.
Hajer, M., & Versteeg, W. (2005). A decade of discourse analysis of environmental politics: Achievements, challenges, perspectives. *Journal of Environmental Policy & Planning, 7,* 175–184.
Hansen, A. (Ed.). (1993). *The mass media and environmental issues.* Leicester: Leicester University Press.
Howarth, D. (2000). *Discourse.* Buckingham: Open University Press.
Jacobs, M. (1999). Sustainable development as a contested concept. In A. Dobson (Ed.), *Fairness and futurity: Essays on environmental sustainability and social justice* (pp. 21–45). Oxford: Oxford University Press.
Jänicke, M., & Jörgens, H. (2000). Strategic environmental planning and uncertainty: A cross-national comparison of green plans in industrialized countries. *Policy Studies Journal, 28*(3), 612–632.
Jonas, H. (1987). Symbolic interactionism. In A. Giddens & J. Turner (Eds.), *Social theory today* (pp. 82–115). Stanford: Stanford University Press.
Jørgensen, M. W., & Philipps, L. J. (2002). *Discourse analysis as theory and method.* London: Sage.
Keller, R. (2004). *Diskursforschung. Eine Einführung für Sozialwissenschaftler.* Opladen: Leske+Budrich.
Keller, R. (2005). Analysing discourse. An approach from the sociology of knowledge. *Forum: Qualitative Social Research (FQS), 6*(3), Art 32 [From http://www.qualitative-research.net/fqs].
Keller, R., Hirseland, A., Schneider, W., & Viehöver, W. (Eds.). (2001). *Handbuch sozialwissenschaftliche Diskursanalyse. Band I: Theorien und Methoden.* Opladen: Leske+Budrich.
Kleinhückelkotten, S. (2005). *Suffizienz und Lebensstile: Ansätze für eine milieuorientierte Nachhaltigkeitskommunikation.* Berlin: Berliner Wissenschafts.
Kollmuss, A., & Agyeman, J. (2002). Mind the gap: Why do people act environmentally and what are the barriers to pro-environmental behavior? *Environmental Education Research, 8*(3), 239–260.
Kuckartz, U., & Rheingans-Heintze, A. (2006). *Trends im Umweltbewusstsein. Umweltgerechtigkeit, Lebensqualität und persönliches Engagement.* Wiesbaden: Verlag für Sozialwissenschaften.
Laclau, E., & Mouffe, C. (1985). *Hegemony and socialist strategy.* London: Verso.
Lafferty, W. M., & Meadowcroft, J. (Eds.). (2000). *Implementing sustainable development: Strategies and initiatives in high consumption societies.* Oxford: Oxford University Press.
Luhmann, N. (1989). *Ecological communication.* Chicago: Chicago University Press.
Luhmann, N. (1997). *Die Gesellschaft der Gesellschaft, Band. 1.* Frankfurt a. M.: Suhrkamp.
McManus, P. (1996). Contested terrains: Politics, stories and discourses of sustainability. *Environmental Politics, 5*(1), 48–73.
Middlemiss, L. & Young, W. (2008). Attitudes are not enough. The importance of context in sustainable consumption. In *Proceedings: Sustainable Consumption and Production: Framework for Action. 2nd Conference of the Sustainable Consumption Research Exchange Network (SCORE!)* (pp. 59–72), 10–11 March, Brussels, Session III–IV.
Neuzil, M., & Kovarik, W. (1996). *Mass media & environmental conflict: America's green crusades.* London: Sage.
Phillips, N., & Hardy, C. (2002). *Discourse analysis: Investigating processes of social construction.* Thousand Oaks: Sage.
Poferl, A., Schilling, K., & Brand, K.-W. (1997). *Umweltbewusstsein und Alltagshandeln.* Opladen: Leske+Budrich.
Reusswig, F. (1994). *Lebensstile und Ökologie. Sozial-ökologische Arbeitspapiere 43.* Frankfurt am Main: ISOE.
Rink, D. (Ed.). (2002). *Lebensstile und Nachhaltigkeit. Konzepte, Befunde und Potentiale.* Opladen: Leske+Budrich.

Rootes, C. (2007). *Environmental protest in Western Europe*. Oxford, New York: Oxford University Press.
Sachs, W. (1997). Sustainable development. In M. Redclift & G. Woodgate (Eds.), *The international handbook of environmental sociology* (pp. 71–82). Cheltenham: Edward Elgar.
Sachs, W., Loske, R., & Linz, M. (Eds.). (1998). *Greening the North: A postindustrial blueprint for ecology and equity*. London: Zed Books.
Schultz, I., & Stieß, I. (2008). Linking sustainable consumption to everyday life. A social-ecological approach to consumption research. In A. Tukker, M. Charter, & C. Vezzoli (Eds.), *Perspectives on radical changes to sustainable consumption and production. System innovation for sustainability 1* (pp. 288–300). Sheffield (UK): Greenleaf Publishing Ltd.
Schulze, G. (1992). *Die Erlebnisgesellschaft. Kultursoziologie der Gegenwart*. Frankfurt/New York: Campus.
Schwenk, O. (Ed.). (1996). *Lebensstil zwischen Sozialstrukturanalyse und Kulturwissenschaft*. Opladen: Leske+Budrich.
Selin, H. (Ed.). (2003). *Nature across cultures: Views of nature and the environment in non-western cultures*. Dordrecht: Kluwer Academic.
Shove, E. (2003). *Comfort, cleanliness and convenience: The social organisation of normality*. Berg: London.
Shove, E., & Warde, A. (2002). Inconspicuous consumption: The sociology of consumption, lifestyles and the environment. In R. Dunlap, F. Buttel, P. Dickens, & A. Gijswijt (Eds.), *Sociological theory and the environment: Classic foundations, contemporary insights* (pp. 230–251). Lanham, MA: Rowman & Littlefield.
Southerton, D. H., Chappells, H., & van Vliet, B. (2004). *Sustainable consumption: The implications of changing infrastructures of provision*. Cheltenham: Edward Elgar.
Spangenberg, J. (Ed.). (2003). *Vision 2030. Arbeit, Umwelt, Gerechtigkeit – Strategien für ein zukunftsfähiges Deutschland*. München: Oekom.
Stahmer, C., & Schaffer, A. (Eds.). (2006). *Halbtagsgesellschaft. Konkrete Utopie für eine nachhaltige Gesellschaft*. Baden-Baden: Nomos.
Umweltbundesamt (Ed.) (2009). *Umweltbewusstsein und Umweltverhalten der sozialen Milieus in Deutschland*. Berlin [From http://www.umweltdaten.de/publikationen/fpdf-l/3871.pdf].
van Kasteren, Y. (2008). What are the drivers of environmentally sustainable consumer behaviour? In *Proceedings: Sustainable Consumption and Production: Framework for Action. 2nd Conference of the Sustainable Consumption Research Exchange (SCORE!) Network* (pp. 173–192), Brussels, 10–11 March, Session III–IV.
Vester, M., von Oertzen, P., Geiling, H., Hermann, T., & Müller, D. (1993). *Soziale Milieus im gesellschaftlichen Strukturwandel. Zwischen Integration und Ausgrenzung*. Köln: Bund.
Vester, M., Hofmann, M., & Zierke, I. (Eds.). (1995). *Soziale Milieus in Ostdeutschland. Gesellschaftliche Strukturen zwischen Zerfall und Neubildung*. Köln: Bund.
Wenzel, E., Rauch, Ch, & Kirig, A. (2007). *Zielgruppe LOHAS. Wie der grüne Lifestyle die Märkte erobert*. Kelkheim: Zukunftsinstitut.

Chapter 6
Psychological Aspects of Sustainability Communication

Lenelis Kruse

Abstract A psychological view of sustainability communication opens up three perspectives. First, it deals with the social and societal construction of complex concepts like 'environment', 'nature' or 'sustainable development', which is realized through both direct and mediated communication; second it analyses (global) human-environment problems and their systemic interrelations hips, which elude immediate sensory perception and depend on visual and verbal communication; and, finally, it focuses on communication, which is an important tool to stimulate mankind to adopt sustainable behaviour patterns.

Keywords Environmental psychology • Sustainable behaviour • Perception of global environmental change • Gap between awareness and action

Environment, Nature and Sustainable Development as Social and Cultural Constructs

From the viewpoint of psychology – that is, environmental psychology – the problem is how to influence and modify non-sustainable behaviour patterns together with those factors on which they are based, such as values, attitudes, knowledge, motivation, habits, social norms, as well as the structural or contextual conditions of such behaviour. From this broad definition of a psychology focused on issues of sustainability, it becomes clear that such program of behavioural change includes and requires much more than communication alone. It also becomes clear that psychology alone cannot accomplish this. A long list of other human science disciplines

---

L. Kruse (✉)
Psychological Institute, University of Heidelberg, Heidelberg, Germany
e-mail: Lenelis.Kruse@psychologie.uni-heidelberg.de

dealing with 'human dimensions of global change' would have to be included, all of which could be summarized under the label of 'human ecology' (Kruse 2004). Typically each sustainability problem also includes aspects that involve natural science. As a result sustainable development requires multidisciplinary – or better yet – interdisciplinary cooperation between natural and human sciences, in which each of the participating disciplines must present, negotiate and integrate their theoretical concepts, their methodologies, and their problem-solving approaches in order to create a scientific basis for the societal process of sustainable development.

There are few analyses of environmental, or rather ecological communication, that miss the opportunity to quote Niklas Luhmann that there can only be a socially shared perception of environmental and of ecological risks if it is communicated (1989). The manner of communication becomes apparent – as, for example, environmental discourse – when certain issues and events are linked to concepts and corresponding valuations. These are created, stabilized or changed through face-to-face interactions or through the media, in scientific and in political discussions, that is, they are socially constructed. The environmental discourse that attracts attention through its large vocabulary of crises and risks, and at the same time of reassurance and alarm, is part of a continually changing social representation that is shared collectively or only by specific groups (Farr and Moscovici 1984; Graumann and Kruse 1990). The concept of 'sustainable development' has not quite reached the status of a social representation; at best, one could speak of a group-specific representation. When a biannual opinion poll on 'environmental awareness in Germany' in 2004 showed that about one third of all interviewees had at least heard of the term 'sustainable development', many saw this as a success (Kuckartz and Rheingans-Heintze 2004), but the very concept was discarded from later polls and replaced by concepts specifying crucial issues of sustainable development, such as intergenerational equity etc. (Umweltbundesamt 2009).

Environmental discourses and societal constructions of the environment often show great cultural variations (Douglas and Wildavsky 1982), not only between distant countries, such as those in the industrialized North and the emerging nations of the South, but also between neighbouring countries. A pertinent, and for some time politically controversial, example was the culturally divergent concept, valuation and use of *Waldsterben* (the 'death of the forests') in Germany and France. The adoption of the German term *le Waldsterben* in French served as a kind of 'distancing function' and reflected the low relevance of this environmental problem in France.

If everyday behaviour patterns are to be changed, it is important to consider group and subgroup-specific constructions and mentalities, which are discussed below under the headings of lifestyles and social milieus.

Perception and Evaluation of Global Environmental Changes

Social representations of the environment, of nature or of sustainability – as substantiated in societal discourse – play a crucial role in gaining attention to the structures and processes needing to be sustainably transformed, with the perception and evaluation of underlying problems being of special relevance. An important catalyst for the conception and dissemination of the principle of sustainable development has been the growing recognition of and concern about the anthropogenic nature of environmental changes, which are based on non-sustainable or 'maladaptive' behaviours of humans towards life-supporting natural resources. The development of an adequate concept of sustainable development requires that humans be seen in their triple role: as causal agents, as victims and – most importantly – as change agents. The requirements for developing the learning processes and competencies of people (as individuals as well as members of groups and social collectives) are considerable, while the structures and processes of human-environment interactions show characteristics that compound the difficulties of learning such competencies (Pawlik 1991; Kruse 1995; Lantermann 2000):

- People lack the requisite sense organs for detecting many environmental conditions and changes, e.g. the ozone hole or radioactive fallout cannot be seen, heard, or smelt. Other changes are so minimal or gradual that they fall below the threshold of 'just noticeable differences'.
- Some human activities have immediate and direct effects on the environment, while others have delayed effects that may not immediately be seen as direct causes of environmental change. In addition to the *time lag* between interference and effect, there is a *spatial* factor that must be considered. For example, the CFC emissions of industrialized nations in the North first developed their harmful effects (depletion of the ozone layer) in the southern hemisphere. This temporal and spatial distance is often accompanied by a social distance between those causing and those affected by environmental deterioration or hazards. The inhabitants of wealthy countries, where pollution often originates, may not realize its effects on a highly vulnerable population in emerging countries, which has few resources to cope with the damages. With global environmental problems it is essential to consider both long-term and long-distance effects.
- Other cognitions come into play when individual effects are very small. This holds true not only for harmful activities but also for many positive behavioural contributions as well (e.g. reduced driving of a private car). Small damages to the environment or improvements are seen as a 'drop in the bucket' and the growing 'stream' accumulating over time is overlooked, as is the dissemination of new behaviour patterns to larger groups.
- In general high complexity, network structures, high dynamics and the non-transparency of human-environment interactions, together with long time horizons and multiply interrelated systems (Dörner 1989) present extreme difficulties for human cognitive abilities. In addition, one has to take into account the restricted or generally unpredictable nature of global developments, which require action under conditions of uncertainty and the development of entirely new decision-making processes and responsibilities (Lantermann 2000).

The invisibility and remoteness from experience of many environmental problems, as well as the inability to perceive correlations between cause and effect, has a number of psychological consequences:

- Where immediate experience is missing it is replaced by indirect experience. On one hand, individuals seek a better understanding through interpersonal communication, which offers social support, especially in cases where the 'reality' cannot be tested. On the other, the mass media assume significant relevance as they transform unnoticeable and abstract facts into images and computer simulations, as they use language to frame problems, thus making them comprehensible. The media thus have a specific role in the social construction of global environmental change. Furthermore, controversial expert debates in the media deserve special mention as they produce 'second-hand non-experience' for the public (Beck 1992).
- In order to make conspicuous and incomprehensible phenomena understandable, individuals will attempt to find a cause, even if a monocausal explanation does not do justice to the complex circumstances, such as the process of climate change (e.g. an accumulation of extreme weather events is seen as a consequence of climate change). Other cognitive strategies that are often regarded as leading to 'errors' in human information processing, but should rather be taken as 'rules of thumb', are the so-called judgmental heuristics. These simplify complex problem-solving processes, but are mostly used in an unreflected fashion (Kahneman et al. 1982). Such judgmental heuristics focus on, for example, the 'representativeness' of information, or cognitive 'availability' or 'framing' the specific presentation of facts. The importance of events that may indeed occur incidentally, like a very hot summer or a surprisingly long winter, may thus be overestimated and taken as an indicator for global warming (representativeness heuristics). The significance if novel or spectacular, picturesque and impressive incidents with great media coverage (dying seals or bird flu) will also be overestimated (availability heuristics).

Research on cognitive strategies and 'biased' findings are of special importance when applied to the appraisal, communication and acceptance of risks.

Moving toward sustainability involves transforming non-sustainable behaviour in many areas of everyday life, such as food consumption or recreational mobility. Ultimately it is all about complex processes of 'un-learning' non-sustainable behaviour patterns and adopting more sustainable ones or, more comprehensively, lifestyles. It also includes the acquisition of decision-making and action-taking competencies that take into account the three dimensions of sustainability, i.e. the environmental, economic and social (Kaufmann-Hayoz and Gutscher 2001). An important condition for this is knowledge about the conceptual foundations, methodologies and instruments of strategies for behavioural change.

The Gap Between Environmental Awareness and Action

In the public, but also in many political discussions, there is a widespread belief that an increase in knowledge and/or strengthening of attitudes will lead – almost automatically – to more sustainable behaviour. As a central instrument, communication is primarily used in the sense of providing one-way information, such as leaflets, professional literature, lectures, radio and television broadcasts. On the other hand, however, there are constant complaints about the 'gap between knowledge and action'. Without being able to give a full account of these seemingly contradictory arguments (Diekmann and Preisendörfer 1992; de Haan and Kuckartz 1996; Kruse 2002), one can conclude that behaviour relevant to the environment and sustainability is influenced by a number of determinants that could be seen as either behavioural barriers hindering sustainable behaviour or as support for non-sustainable behaviour.

Since the 1970s a large body of research has been undertaken in the field of environmental awareness and action (e.g. Gardner and Stern 2002; Gifford 2007a; Schmuck and Schultz 2002) in order to understand the problems of sustainability learning. In the following the focus will be on 'environmentally relevant' or 'pro-environmental' learning. It should be noted however that there is still a need for more painstaking research into sustainable development, especially in view of its spatio-temporal, and global aspects, of its relationship to intergenerational justice and responsibilities as well as of the need for promoting sustainable behaviour patterns (for an overview, see APA 2010).

Multiple Determinants of Environmentally Relevant and Sustainable Behaviour

In response to an increasing interest in the everyday psychological problem of 'environmental awareness' or 'environmental concern', psychology has treated 'environmental awareness' as a scientific concept, in addition to examining other determinants of pro-environmental or conservation behaviour. Several explanatory models have been developed and empirically tested. The focus is on finding intervention strategies and instruments to modify non-sustainable behaviours and to promote more sustainable behaviour patterns. It is important to carefully evaluate these instruments as to their effectiveness and efficiency in various contexts of action.

Knowledge alone is not a guarantee for pro-environmental behaviour, especially abstract knowledge about environmental problems, which lacks an action orientation and is almost invariably based on survey questionnaires or public opinion polls. Knowledge, however, is one of the necessary factors that has to be taken into account and more recent research has made attempts to specify knowledge areas in a much more concrete, i.e. action-specific, manner, and furthermore to differentiate between types of knowledge, such as systemic knowledge, action knowledge and prognostic or effective knowledge, all of which will more closely correlate with concrete action. And, of course, in order to understand how knowledge is acquired, it is important to study aspects of communication, such as how factual information is actually presented.

There are further factors to be taken into account if pro-environmental and sustainable behaviour is to be promoted. These factors may be classified as individual, interpersonal/social and external/structural conditions.

- Aside from problems of knowledge, *individual factors* include problems of the perceptibility of environmental conditions and changes, as well as risk construction, understanding complex systems and the accompanying processes of information processing. Further individual factors include value orientations and attitudes as well as personality characteristics or habitual motives (e.g. egocentrism, altruism or social responsibility), but also temporary emotions like fear of failure or hope for success when pro-environmental actions are at stake.
- Social norms and values of membership and reference groups are examples of *interpersonal* and *social* factors. Values of a society as a whole (for example, orientation toward the principle of sustainability) are important, as are social, economic, political and cultural norms that are conveyed and filtered through the mass media. Social interaction and communication play an important role as they may facilitate or impede certain activities, and observation of others' behaviour (social models) usually has a strong influence on one's own behaviour. Furthermore, existing social networks (neighbourhoods, teams at school or at work) should be taken into consideration as they can facilitate the process of participation and learning. Another important aspect of a social situation is conflict among interest groups, in which supposed winners and losers of a specific action taken (e.g., reducing the speed limit in a residential area) may contribute to completely divergent perspectives and appraisals of a controversy.
- *External structures and contexts* can advance or hinder sustainable actions. There is often a lack of opportunities for action (e.g. lack of availability of public transportation or energy-saving devices) that are necessary for resource-saving behaviour. Another aspect of external structures are the various incentives for positive behaviour, with monetary rewards (eco-tickets, subsidies for solar panels) being most important, but also non-monetary rewards, such as social recognition or public praise having some influence.

For interventions to be successful the entire context of ecological and socio-cultural conditions (climate, resource availability, economic, legal, technological and scientific educational opportunities) has to be taken into account.

Strategies and Instruments for Promoting Sustainable Behaviour

There are a great variety of explanatory models and strategies about how behaviour can be made more sustainable. Environmental psychology has developed quite a number of intervention strategies to enhance environmental awareness as well as increase the likelihood of undertaking environmentally relevant actions (e.g. Gardner and Stern 2002; Gifford 2007a). In the meantime the perspective has been broadened to address more complex patterns of awareness and actions, such as climate change and sustainable development (e.g. APA 2010; Gifford 2007b). The various intervention methods can be roughly classified into cognitive and behavioural strategies. The latter can be subdivided into antecedent measures preceding critical behaviour and consequence measures following critical behaviour.

Cognitive strategies try to influence cognition and knowledge of environmental conditions and changes by working with information and educational approaches (therefore, they are often summarized under the label of 'education'). In this context issues of information presentation, communication media, but also the characteristics of the communicator and the recipients are of special significance. However, more effective than pure information is concrete feedback about individual success and failure as well as learning from models – an example of an antecedent strategy. Other examples include prompts (e.g. signs or posters), self-defined or adopted goals and private or public commitments. Consequence measures, which as a rule are less effective than antecedent ones, mainly work with reward and punishment, but also with individual or collective feedback.

In general, it can be said that a combination of intervention instruments will only be successful in promoting sustainable behaviour if it takes into consideration specific target groups (e.g. car drivers, nature conservationists or tourists), fields of action (e.g. mobility or conservation of nature) and specific contexts (workplace, place of vacation or suburban dwellings). A fundamental condition, found in applied research projects, for advancing sustainable development in specific contexts is the evaluation of measures (e.g. Dwyer et al. 1993).

Information and Communication

Almost all interventions make use of information and communication. If the emphasis lies on cognitive or education-oriented intervention, then the focus is on various kinds of information materials. Social science research however has often confirmed that information alone is hardly ever effective in changing behaviour. Their effectiveness would improve if the most important principles of information and communication would be taken into account.

Classic communication models involve analysing a number of specific components:

- *Who* is the communicator? Competence and credibility are important. In addition to personal appearance prestige and affiliation with an organization are important.
- *What* is communicated? This addresses the issue of information content and design. Attitude and behaviour changes are more likely if there is information that is accurate, easily understandable, personalized and vividly presented. It should link to existing beliefs, interests and the knowledge of the recipients, or target groups, so that it is able to attract attention and can be understood. Since the presentation of facts can also be used to evoke emotions (joy, fear etc.), it is useful to consider the research findings on the effects of emotions on attitude change or behaviour modification.
- What is the *intention or function* of a communication situation? Kaufmann-Hayoz and Gutscher (2001) suggest a useful distinction between communication instruments *without* direct request or *with* direct request. The first type of communication presents facts, options, standards and objectives as well as model behaviour or

feedback with no intent to persuade or make calls for action, whereas communication with direct request is meant to convince individuals about facts, goals and norms, present reminders, send appeals and encourage self-commitment.
- What *media* is used? The choice is dependent both on the purpose of communication and on the target of communication. It must be clarified, for example, whether target persons need to be addressed individually or whether interpersonal exchange is to be stimulated for the purpose of fostering participatory processes. Also, it has to be decided what type of media and media design will be successful to attract attention, stimulate further information seeking or increase knowledge about the functioning of complex systems.
- What is the desired *success* of communication? This is a necessary, though a sensitive issue. Is it enough for a problem to be simply discussed, or is the intention rather to gain noticeable long-term behavioural change? What is the relation of the financial investment to the observable effects?

From the perspective of psychological intervention research and practice, well-designed information and communication processes are a necessary but not sufficient condition to promote the sustainable development of society. Even if it seems that newer strategies in environmental protection and the sustainability movement, such as participation, moderation and mediation or social marketing, put much emphasis on communication, without the introduction and design of additional factors, especially the provision of incentives for action and adequate opportunities, there will be no sustainable development that is also undertaken by the concrete actions of individuals.

Sustainable development implies a continual process of changing human-environment interactions, a process that must repeatedly focus on new objectives that result from the interdependencies between ecological, economic, social and cultural conditions. It is a global process that must be implemented internationally, nationally, regionally and locally, as well as at all levels of societal organization. Psychology, specifically environmental psychology, can contribute its concepts, methodologies and research findings about the various modes of human-environment interactions and can thus support learning processes for sustainable action. Communication of and about sustainability in society must prepare the ground for the multiple and multidisciplinary use of strategies and interventions to move people towards sustainable lifestyles and behaviour.

References

APA (American Psychological Association) (2010). *Psychology and global climate change: Addressing a multi-faceted phenomenon and set of challenge*s. A Report by the American Psychological Association's Task Force on the Interface Between Psychology and Global Climate Change. Retrieved July 30, 2010, from www.apa.org/science/about/publications/climate-change.pdf.

Beck, U. (1992). *Risk society: Towards a new modernity*. Newbury Park: Sage.

de Haan, G., & Kuckartz, U. (1996). *Umweltbewusstsein. Denken und Handeln in Umweltkrisen.* Opladen: Leske + Budrich.

Diekmann, A., & Preisendörfer, P. (1992). Persönliches Umweltverhalten. Diskrepanzen zwischen Anspruch und Wirklichkeit. *Kölner Zeitschrift für Soziologie und Sozialpsychologie, 44,* 226–251.

Dörner, D. (1989). *Die Logik des Misslingens.* Reinbek: Rowohlt.

Douglas, M., & Wildavsky, A. (1982). *Risk and culture.* Berkeley: University of California Press.

Dwyer, W. O., Leeming, F. C., Cobern, M. K., Porter, B. E., & Jackson, J. M. (1993). Critical review of behavioural interventions to preserve the environment. *Environment and Behaviour, 25,* 275–321.

Farr, R., & Moscovici, S. (Eds.). (1984). *Social representations.* Cambridge: Cambridge University Press.

Gardner, G., & Stern, P. (2002). *Environmental problems and human behaviour.* Boston, MA: Pearson Custom Publishing.

Gifford, R. (2007a). *Environmental psychology: Principles and practice.* Colville, WA: Optimal Books.

Gifford, R. (2007b). Environmental psychology and sustainable development: Expansion, maturation, and challenges. *Journal of Social Issues, 63,* 199–212.

Graumann, C. F. & Kruse, L. (1990). The environment: Social construction and psychological problems. In: H. Himmelweit & G. Gaskell (Eds.), *Societal psychology* (pp. 212–229). Newbury Park: Sage.

Kahneman, D., Slovic, P., & Tversky, A. (Eds.). (1982). *Judgment under uncertainty: Heuristics and biases.* Cambridge: Cambridge University Press.

Kaufmann-Hayoz, R., & Gutscher, H. (Eds.). (2001). *Changing things – moving people: Strategies for promoting sustainable development at the local level.* Basel: Birkhaeuser.

Kruse, L. (1995). Globale Umweltveränderungen: Eine Herausforderung für die Psychologie. *Psychologische Rundschau, 46,* 81–92.

Kruse, L. (2002). Umweltverhalten – Handeln wider besseres Wissen? In G. Hempel & M. Schulz-Baldes (Eds.), *Nachhaltigkeit und globaler Wandel* (pp. 175–192). Frankfurt: Peter Lang.

Kruse, L. (2004). Umweltpsychologie als Humanökologie. In W. Serbser (Ed.), *Humanökologie. Ursprünge – Trends – Zukünfte* (pp. 270–293). München: Oekom.

Kuckartz, U., & Rheingans-Heintze, A. (2004). *Umweltbewusstsein 2004. Ergebnisse einer repräsentativen Bevölkerungsumfrage.* Berlin: Bundesministerium für Umwelt, Naturschutz und Reaktorsicherheit.

Lantermann, E. D. (2000). Der globale Wandel als Herausforderung der Umweltbildung. In A. Grewer, E. Knödler-Bunte, K. Pape, & A. Vogel (Eds.), *Umweltkommunikation. Öffentlichkeitsarbeit und Umweltbildung in Großschutzgebieten* (pp. 73–81). Berlin: PR Kolleg Berlin.

Luhmann, N. (1989). *Ecological communication.* Cambridge: Polity Press.

Pawlik, K. (1991). The psychological dimensions of global change. *International Journal of Psychology, 26*(5) Special Issue, 545–673.

Schmuck, P., & Schultz, W. P. (2002). *Psychology of sustainable development.* Boston: Kluwer.

Umweltbundesamt (Ed.) (2009). *Umweltbewusstsein der sozialen Milieus in Deutschland. Repräsentativumfrage zum Umweltbewusstsein und Umweltverhalten im Jahre 2008.* Berlin.

Chapter 7
Media Theory and Sustainability Communication

Claudia de Witt

Abstract Since communication about sustainable development takes place in a mediatised knowledge society it seems appropriate to investigate the importance of the media to achieve this goal. The question then arises as to what contribution the media as communication media can make to the diffusion of awareness about sustainability and to what extent they can influence and promote social discourse. Especially the new media, such as Web 2.0, seem to have the potential to transport 'information' through global communication networks across national borders and through participatory processes to ensure involvement in global discourses about sustainability communication.

Keywords Media theory • Web 2.0 • Effect and use analysis • Media communication • Global communication

Sustainability, Communication and the Media

Communication is considered a means of anchoring the vision of sustainable development in society. In general, successful communication involves, according to Schmidt (1993), a mastery of language, a mutual ability and willingness to communicate and knowing which discourse the communication act is a part of. Furthermore, it involves accounting for the "social structures of a communicative situation in order to be able to assess the allocation of roles in communication" (Schmidt 1993: 109). The first physical-technological use of the term communication was made by Stephen Gray (1729) in connection with his discovery of what he

C. de Witt (✉)
Institute of Educational Science and Media Research, FernUniversität, Hagen, Germany
e-mail: claudia.dewitt@fernuni-hagen.de

called a 'communication thread'. In his electrostatic experiments he stretched a hemp thread and attached it to a battery, with his assistants keeping the thread damp and signalling when the 'electric virtue' arrived. This 'communication thread' anticipated the electronic transmission of data, which in the form of the telephone and internet has become an indispensable part of our lives today.

This is an example of how new media change communication processes. Media science researchers do not investigate a particular medium or the effects of particular medium, but the reciprocal effect of media on our perception of reality and on life in society. McLuhan (1964) has shown persuasively that it is the media themselves and not the content they transport that need to be the focus of research. Media do not – as was once the assumption – depict reality. Instead they create reality. While Pross (1972) differentiates primary, secondary and tertiary media according to their channel of communication, Luhmann (1997) for example sees media as symbolically generalised media of communication, which include money, love, truth, power and value. From this perspective media are anything that mediates – and they do not need to be related to a technology or even communication itself. As a means of communication however media are also instruments that serve to propagate messages. Their functions in talk can be found in the focusing on particular topics, enabling opinions to be formed, criticism to be made or control to be exercised. And this is exactly why they are so interesting for a larger perspective on communication about sustainability.

There is an international discussion among researchers to demonstrate theoretically the enormous potential of new media to encourage people to become responsible actors, not only through the contents transmitted by media but also through the democratic communities they initiate. At the same time global communication networks are moving public social discourse from the level of national to global debates (Castells 2008; Bekkhus and Zacchetti 2010). There are historical and theoretical analyses of the relationships of media, humans and communication (e.g. Leverette 2003; Davidson 2009; Rantanen and Downing 2009; Buck et al. 2010).

Theoretical Perspectives on Media

Media Revolution as a Subject of Media Theory

Especially in revolutionary periods in communication and media cultures, there is a need for theoretical reflection. The emergence of new media technologies, through socio-cultural changes or paradigm shifts in society, science or culture lead to uncertainties, to the loss of competencies, to new concepts and re-evaluations (Rusch 2002). Revolutions in media are "upheavals, which in similar media, photography and film, took place over a century ago and today in digital form are redefining our life" (Käuser 2002: 256). However new media are also characterised by the reversal, recycling and reinforcing of older media elements. The new develops against a background of the old. This process is impressively illustrated by the concept of the 'tetrade' (McLuhan and Powers 1995). Technological innovations do not imply a break with what is now, but on the contrary show a continuous development.

Functions and Approaches of Media Theories

Media theories attempt to "reflect and clarify the identity, functions and status etc. of media in society and for the individual" (Rusch 2002: 252). They describe, explain, criticise or shape the means of communication and reception while referring back to the conditions of their use. These include the technological, cognitive, social and cultural conditions, effects and consequences. Media theories are cited, for example by Maletzke (1998) as being the most important theories in debates about cultural and historical theories. Media theories are often concerned with the fundamental question of what effects media have on social life, of how media effect our perception of the world. These theories can therefore be an important factor in communication finding its place in social discourse. There is then not just one media theory but a number of different theoretical ways of accessing an understanding of the effectiveness and use of media. With a view to the increasing networking and concentration of media and media content, to the increasing intermediality, Leschke (2007) sees the necessity to convert media theories into 'form theories'. He justifies this paradigm shift with the observation that the orientation towards a single media is obsolete and instead exchange processes between media, especially regarding their forms, has become commonplace: "that a medium with its forms remains essentially isolated is practically unthinkable. On the contrary, such medial forms as forms of games, narrative forms and the organisational forms of hypertext and persuasive communication all circulate through the media system and through the media. At the same time it is largely irrelevant where they come from and what their ontological quality is. It is these medial forms that create the network among the media on the level of their products" (Leschke 2007: 5; also Leschke 2010).

Media in the Paradigm of Systems Theory

Media are "no longer understood as merely techniques of communication, as instruments for the diffusion and storage of information, but more as instances of selection and interpretation that actively intervene in the social construction of reality" (Maletzke 1998: 124). Systems are sets of elements in reciprocal relationships. They are constituted by their boundaries to the environment and their relationships to this circumscribed environment. System boundaries separating the system and the environment are constituted through the differences of these relationships.

In recent developments in systems theory there are efforts to overcome the traditional analytic isolation of individual systems and to understand a system through its connections with its environment. However, it must be added that the systems approach is often considered to be "formal, abstract and empty" and is not generally accepted (Maletzke 1998: 132).

Media Theory from the Perspective of Constructivism

Constructivism is a different approach. The subject is seen as central, as an active self-referential being that from the "material that his senses provide him with actively builds a world through selection, projection, signification and interpretation; he constructs his world, and in a way that is unique and individual, though admittedly one that is also shaped by social and cultural conditions" (Maletzke 1998: 126). Constructivist media theories see media then not only as technological institutions that send messages or transport information but as systems offering models or designs for reality, which are constructed by autopoietic systems for autopoietic systems (Schmidt 1994). Both the media world as well as the real world are merely constructions of human beings. Constructivism on the one hand postulates that media produce events and structure reality without the recipient knowing where the structuring elements are. The new constructivism paradigm also entails new media reception. Media are seen as having a triggering function. Media information is processed by the recipient in accordance with his prior knowledge and cognitive system. "The content offered by media cannot be considered a depiction of reality for a number of reasons. It is content that triggers cognitive and communicative systems to initiate a construction of reality within their respective systemic conditions. If this content is not made use of, then the media transports nothing at all" (Schmidt 1994: 8). For Schmidt then the "constructivist interest in media is focused on the following question: what role do the media play for the construction of reality and the culture of a society" (1998: 37)?

Media in Critical Theory

There are a number of groupings of critical theories, all of which agree on the basic issues but differ in their detail (Maletzke 1998). A general characteristic of critical approaches is their social-political orientation. Based on the work of the Frankfurt School, social relationships are seen as power relationships, as structures and processes that are not compatible with ideas of equality and democracy and must therefore be changed. Media are seen as a part of the culture industry, which follows a capitalistic logic in its efforts to influence people. Media are studied in the context of ownership relationships, conditions of production and individual participation. As understood by critical-social theoretical approaches (including

Theodor Adorno, Max Horkheimer, Hans Magnus Enzensberger), media have three functions:

- persuasive function: media serve "to integrate people in the system of consumption without arousing their resistance"
- orienting function: media influence the consciousness of people "by their biased tendencies in representing the options for action"
- ontological function: media determine "through the conservative archiving of social knowledge the consciousness people have of themselves as a species" (Viehoff 2002: 227).

This culturally pessimistic view assumes that the communicative power of the media "destroys the actual emancipatory functions of social communication, especially however those of the aesthetic form of communication. This tendency can only be broken by negative critique. This makes the 'social subject' (…) appear as a more or less passive object of communication organised by the media" (Viehoff 2002: 227). This position has to be seen in the context of its origin, which has changed since the 1970s and 1980s, when the emancipatory functions of mass medial cultural communication were increasingly discussed. Cultural communication had more of a positive effect on the development of social identity. Communication and cultural sociological concepts (e.g. those of Hans Jonas, Pierre Bourdieu, Claus Offe) see in medial communication the potential for class-specific interpretation and cultural identity (Maletzke 1998: 227ff.). In this view both in the perception and the understanding of media content the individual cognitive efforts of the recipient(s) are the basis for comprehensive explanations of media reception (Viehoff 2002).

The view of Adorno and Horkheimer, according to Kloock, "with its biting critique of the modern" (Kloock 2003: 24) is today an isolated one. Although the effects of mass production and industrialisation support their radical theses, the contamination of our food for example has not led to fundamental change. By contrast the media-conditioned change of the world in regard to time and place is still of crucial importance – and not only for media theory research. In particular increasing globalisation leads especially to changes in space and time coordinates. "Much that is place-bound will disappear" (Kloock 2003: 24). Since the 1990s there has been a paradigm shift from the 'dialectic of enlightenment' to the observation that in the wake of increasing digitalisation and the global interconnectedness of all media new social patterns and a new understanding of reality in the context of communication and media is being discussed.

Effect and Use Analysis

Media create topical interest, assessments and models for public opinion. In the context of the 1972 American presidential election, McCombs and Shaw (1972) put forward the thesis that the media had an agenda-setting function that influenced which topics people talk about. From this perspective, mass-communication messages precede opinion-makers. The effectiveness of the agenda-setting effect depends on the urgency of the topic. Topics that are directly experienced are communicated less often than topics that are not personally experienced. The effect is also dependent on the type of medium. While topics broadcast on television tend to have a short-term effect, print media are more likely to have a long-term one.

The communication of topics through the media has varying degrees of duration, which can be shown using a number of different models (Schenk 1987):

- Cumulative model: an intensification of the reporting leads to a higher ranking of the topic on the audience agenda.
- Threshold model: a topic becomes part of the audience agenda when a minimum amount of reporting has taken place.
- Inertia model: when a topic has achieved a certain level of importance on the audience agenda then increases in that importance through more intensive reporting are unlikely to occur.

Media effect approaches such as agenda setting have long played an important role in media science discourse. While they dealt with the question of what media do with people, use approaches contradict the idea that media have a control function and that recipients are only passive. The uses-and-gratification approach describes people as recipients who look for the satisfaction of their needs in the media. "Media use in the form of selection and attention follows the principle of the use that the recipient expects from it" (Maletzke 1998: 119). An individual intervenes, from this perspective, in the process of media communication by "selecting, testing and rejecting media content; and often enough he resists media content" (Maletzke 1998: 119). The knowledge that humans have a distance to things, that they define their situation and see their perception and experiences not as passive reception may not be new (Maletzke 1998: 122), nevertheless it is important that both effect and use approaches are taken into account.

Visual anthropology is also concerned with the constitution of culture by the media. It describes how media intervene in cultural perspectives and which effects their technological implications have in cultural use. Especially cultural studies precede from the assumption that types of reception are determined by a given cultural tradition while at the same time they are continually changed by the particular structures of the different media. According to Rusch, media theory has its historical beginning "as criticism of the written form in Plato's Phaedo" (2002: 253). "Plato criticises the use of writing as a new medium that – indirectly – weakens memory but especially that it leads to problems of interpretation as those unfamiliar with a topic could read a text and as a result of their lack of knowledge of the topic or their differing experiences are barely able to understand its meaning or what was intended by the author" (see also Mersch 2007).

In retrospect every medium that has been introduced has effects on social and individual communication. Goody (1986) for example notes that historical writing, the bureaucratisation of trade and administration or the establishment of legal regulations are a consequence of literate culture, while for Bertolt Brecht, who was considering the possibilities of the audience to engage in active participation, the new medium of radio was an 'instrument of mass enlightenment and political agitation' (Rusch 2002). The computer as the most important new cultural medium is said to lead to children and young people to lose literacy competency and to an increasing orientation towards entertainment needs instead of information and education. In this sense Enzensberger (1988) argues for his thesis of 'television as a zero medium' and promotes an emancipatory use of media.

The demand for an emancipatory use of mass media, especially heard in the 1980s and 1990s, is joined in the twenty-first century by the participatory use of the internet, including weblogs, wikis, video portals and social online networks such as XING, Facebook, MySpace, studiVZ and so on. These applications allow the user himself to become a producer of information in the World Wide Web and so to take part in the development of global communication processes.

Media 2.0 and Participatory Communication Processes

Scott (2006) points out that there are a large number of challenges facing communication about sustainability research and emphasizes the importance of having good contacts to selected open-minded specialist journalists in all types of media (radio, television, online etc.). The Institute for Sustainable Communication (ISC) is confident in the ability of new media "to increase the understanding of sustainability best practices and to assist individuals and organizations in adopting more sustainable print and digital media workflows aligns with Earth Day" (ISC 17.1.2010).

Siemens has shown interest in the possibilities of Web 2.0 and social software and has developed the idea of networking in a theoretical framework it calls 'connectivism', which is a new theory of learning in the digital era (Siemens 2004, 2006). This concept describes the influence new media have on our form of communication and our form of life. The new media enable infinite connections among people, sources of information, topics and concepts; they produce information and communication networks, which are often however superficial. Such knowledge and communication networks consist of the properties *'diversity'*, *'autonomy'*, *'interactivity'* and *'openness'* (Siemens 2006: 28).

At the same time this concept describes a new generation of media users, who through a changed use of communication, e.g. using mobile devices, blogging or twitter have become 'digital natives', that is individuals belonging to a generation that has grown up and feels at home with new media. Their communication processes generate personal networks and collaborative scenarios that influence the sustainability of communication. Much discourse about new media, about Web 2.0, deals with the influence of weblogs on public opinion and the change in the role of traditional media (Schmidt 2006). Zerfaß and Boelter (2005) speak of 'the new opinion-makers' and assume that the previous rules of public communication are being changed by the building of interpersonal digital networks. The rise of online networks is accompanied by participation in knowledge of the world, and its development, which was unknown in traditional media.

Regarding Web 2.0, Geert Lovink (2007) has a more critical view of internet culture. "It is true that the internet questions authority and power in new ways. The old sources of knowledge and taste are – let us put it carefully – threatened. But, first of all, the decline of the position of the critic is part of the history of the twentieth century, and the network has only accelerated it. Secondly, the need for information of an assured quality is enormous, especially today. Thirdly, the journalistic perspective on the channels of communication distracts from what interests me most on the social web. A new virtual space has been created, one in which I can position myself beyond family, work, business" (Lovink 2007).

It is not the news and opinions in the net that are important, but the special way of representing oneself. A typical example is the blogger. "Blogging sustains a cult of the individual in a situation hostile to individualism and to this extent it is the successor of the diary. However it is a completely new diary culture, one that is neither public nor private, but takes place in an intermediary place. Although blogging is writing, it has something informal to it. Like a rumour it pales and fades very quickly" (Lovink 2007). Nevertheless individual self-expression and opinions are immediately available for reading world-wide. Given these new users and global social movements, it is thus only consequent that media theories should be developed to be more "*cross-disciplinary*" (Ekecrantz 2007: 177), both theoretically (e.g. Rossiter 2003) as well as qualitative (e.g. Gauntlett 2007).

Media and Sustainability Communication

Communication about media is a possibility to further the vision of sustainable development. This chapter is not about which medium is more sustainable (but see here Carli 2009). It is an investigation of media from a number of theoretical perspectives, which should make clear that media have a social orientation function. However communication strategies and processes must be scrutinised from a perspective of medial structures. It can be seen that individual communication as technologically transmitted is gaining increasingly in importance. It can be held that

- media theory would like to explain communication through social conditions
- knowledge about the effects and use of new media enhances participation in changing communication processes
- media communication must take into account both communication culture problems in local, regional and national areas as well as new social ties across national borders, involving new communication culture opportunities as well as problems for humankind
- global communication about new media opens opportunities for individuals to communicate across national borders about how the ecological basis of human life or distributive justice across synchronous and asynchronous communication spaces can best be secured
- media communication has become global communication.

Without doubt part of a strategy of sustainable communication is providing opportunities so that a subjective perception and experience can occur that will favour acceptance of the vision. An appreciation of the problem is closely related to the degree an individual feels affected by the problem. This involves recognising the meaning and importance of these opportunities. Communication through new media can mean expanding one's own life and world. Communication about the life chances of future generations involves responsibility for media generations. Sustainability communication requires then not least an awareness of the socialisation and learning conditions of future media generations, who are at home in an increasingly globalised world with digital networks of communication cultures.

References

Bekkhus, N., & Zacchetti, M. (2010). A European approach to media literacy. *International Journal of learning and Media, 1*(1). Retrieved July 30, 2010, from http://ijlm.net/news/european-approach-media-literacy.

Buck, M., Hartlin, F., & Pfau, S. (2010). *Randgänge der Mediengeschichte*. Wiesbaden: VS.

Carli, D. (2009). *Are you prepared to fight for the future of print media, digital media or both?* Retrieved December 7, 2009, from www.sustainablecommunication.org/resources/articles/88-are-you-prepared-to-fight-for-the-future-of-print-media-digital-media-or-both.

Castells, M. (2008). The new public sphere: global civil society, communication networks, and global governance. *The Annals of the American Academy of Political and Social Science, 616*(1), 78–93.

Davidson, C. N. (2009). Blamed for change: historical lessons on youth, labor, and new media futures. *International Journal of Learning and Media, 1*(3), 11–18.

Ekecrantz, J. (2007). Media and communication studies going global. *Nordicom Review*, Jubilee Issue, 169–181.

Enzensberger, H. M. (1988). Das Nullmedium oder Warum alle Klagen über das Fernsehen gegenstandslos sind. In H. M. Enzensberger (Ed.), *Mittelmaß und Wahn Gesammelte Zerstreuungen* (pp. 89–103). Frankfurt a. M: Suhrkamp.

Gauntlett, D. (2007). *Media studies 2.0*. Retrieved October 27, 2009, from www.theory.org.uk.

Goody, J. (1986). *The logic of writing and the organization of society*. Cambridge: Cambridge University Press.

Institute for Sustainable Communication. *Media*. Retrieved May 05, 2010, from http://www.sustainablecommunication.org/media/news.

Käuser, A. (2002). Medienumbrüche. In H. Schanze (Ed.), *Metzler Lexikon* (pp. 255–257). Stuttgart: Metzler.

Kloock, D. (2003). *Klassiker der Medientheorie*. Duisburg: Studienbrief für das Online-Studienprogramm Educational Media.

Leschke, R. (2007). *Mediale Formen zwischen Intermedialität und Vernetzung*. Retrieved December 7, 2009, from http://www.medienmorphologie.uni-siegen.de/downloads/leschke_intermedialitaet.pdf.

Leschke, R. (2010). *Medien und Formen* (Eine Morphologie der Medien). Konstanz: UVK.

Leverette, M. (2003). *Toward an ecology of understanding: Semiotics, medium theory, and the uses of meaning*. In Image & Narrative, Issue 6. Medium theory, January 2003. Retrieved May 05, 2010, from http://www.imageandnarrative.be/mediumtheory/marcleverette.htm.

Lovink, G. (2007). *Ich blogge, also bin ich. Immer mehr Menschen produzieren sich im Internet. Wir treten ein in die Epoche des 'digitalen Nihilismus'*. Retrieved December 1, 2009, from http://www.zeit.de/2007/52/Interview-Geert-Lovink.

Luhmann, N. (1997). *Die Gesellschaft der Gesellschaft*. Frankfurt a. M: Suhrkamp.
Maletzke, G. (1998). *Kommunikationswissenschaft im Überblick*. Opladen: VS.
McCombs, M. E., & Shaw, D. L. (1972). The Agenda-setting function of mass media. *Public Opinion Quarterly, 36*, 176–187.
McLuhan, M. (1964). *Understanding media: The extensions of man*. New York: McGraw-Hill.
McLuhan, M., & Powers, B. (1995). *The global village* (Der Weg in die Mediengesellschaft in das 21. Jahrhundert). Paderborn: Junfermann.
Mersch, D. (2007). *Medientheorien zur Einführung*. Hamburg: Junius.
Pross, H. (1972). *Medienforschung – film, funk, fernsehen*. Darmstadt: Habel.
Rantanen, T., & Downing, J. (Eds.) (2009). Global media and communication, December 2009, Vol. 5.
Rossiter, N. (2003). *Processual media theory*. Melbourne DAC2003 Paper. Retrieved December 1, 2009, from http://hypertext.rmit.edu.au/dac/papers/Rossiter.pdf.
Rusch, G. (2002). Medientheorie. In H. Schanze (Ed.), *Metzler Lexikon* (pp. 252–255). Stuttgart: Metzler.
Schenk, M. (1987). *Medienwirkungsforschung*. Tübingen: Mohr.
Schmidt, S. J. (1993). Kommunikation-Kognition-Wirklichkeit. In G. Bentele, V. Günter, & M. Rühle (Eds.), *Theorien öffentlicher Kommunikation* (pp. 105–117). München: UVK.
Schmidt, S. J. (1994). Die Wirklichkeit des Beobachters. In K. Merten, S. J. Schmidt, & S. Weischenberg (Eds.), *Die Wirklichkeit der Medien* (p. 8). Opladen: VS.
Schmidt, J. (2006). *Weblogs* (Eine kommunikationssoziologische Studie). München: UVK.
Scott, A. (2006). Communicating sustainability research. Theoretical and practical challenges. In W. L. Filho (Ed.), *Innovation, education and communication for sustainable development* (pp. 535–557). Frankfurt: Lang.
Siemens, G. (2004). *Connectivism: a learning theory for the digital age*. Retrieved November 27, 2009, from http://www.elearnspace.org/Articles/connectivism.htm.
Siemens, G. (2006). *Knowing knowledge*. Retrieved November 27, 2009, from http://elearnspace.org/KnowingKnowledge_LowRes.pdf.
Viehoff, R. (2002). Medienkultur. In H. Schanze (Ed.), *Metzler Lexikon* (pp. 226–229). Stuttgart: Metzler.
Zerfaß, A., & Boelter, D. (2005). *Die neuen Meinungsmacher*. Graz: Nausner & Nausner.

Chapter 8
Communication Theory and Sustainability Discourse

Andreas Ziemann

Abstract Ecological and sustainability discourses are communicative processes. This chapter focuses on communication theory in order to explore the communicative and social aspects of sustainability discourse, in particular reflexivity, commitment and normalisation. Consequences for sustainability communication are discussed.

Keywords Communication theory • Communicative process • Functional differentiation • Characteristics of sustainability discourse • Sustainability communication

The interpretation of 'sustainable development' is as multi-facetted as its strategic realisation. There are few communication theoretical analyses of this term, of its discourse coherence and operationalisation. Research into sustainability and sustainability communication often assign communication a secondary status – as if it were possible to first discuss sustainability, then plan and implement it, and finally communicate it. The opposite is the case. As soon as something has become an issue – and individuals have made a series of specific contributions to that issue – then communication is taking place. Neither sociality nor social structures, neither technology nor ecology are independent of the communication of either given or thinkable situations. It is only through and as communication that an event or an object receives social relevance and meaning. And every event that is well known today has already been through the selection and production machinery of the mass media. The discourse of sustainability is also – as is ecological discourse in general – above all a communicative process event within society. If an awareness of ecological problems and sustainability is not communicated, then it is socially irrelevant, even non-existent.

A. Ziemann (✉)
Bauhaus-Universitaet Weimar, Fakultaet Medien, Bauhausstr. 11, D-99423 Weimar, Germany
e-mail: andreas.ziemann@uni-weimar.de

Sustainability discourse is less about interpersonal contacts and social relationships and more about global living conditions, as well as social values and structure. The foundation for communication theory outlined below is then based not only on general principles but is also embedded in a theory of society.

Communication Theory

In contrast to a techno-scientific understanding of communication, which has yielded a number of complex transmission models (essentially of information transmission between sender and receiver through a given channel), the social and human science description of communication begins with face-to-face contact. Communication is defined as the human and technologically based activity of the reciprocal use of signs and the reciprocal interpretation of signs for the purpose of successful understanding, coordinating action and shaping reality (Krallmann and Ziemann 2001: 13).

Communication is thus a social process in which at least two open-minded, spatially bound actors are involved. With the help of signs, language and symbols – whose effect on themselves and on others the participants observe – social orientation, reciprocal control and informative action take place. The necessity of communication can be found in the human condition: each consciousness is isolated, our neurophysiological, cognitive, emotional processes are mutually unobservable and there is no direct access to the thoughts, attitudes and intentions of the other. It is through communication that 'the interior is exteriorised', that we can inform each other, that we become social creatures. Communication is thus the principle of societal organisation itself.

As a completed event – and in comparison to the attitudes, motives and goals of those involved – communication is then something socially separate – in systems theory we would say that it is 'emergent'. That is why the meaning and effect of communicative events cannot be attributed to one of the participants, nor can they be mentally inferred. Interpersonal sequences of events, relationships, conversations and discourses have an immanent momentum and self-organisation. Out of joint talk and action arises a social event that displays an asymmetric relation between self and other, as a dialectic interrelationship.

Following Luhmann (1995, 1997) we can formulate this more radically. No human subject is the author or transporter of communication and no single consciousness can purposefully order communication. Communication itself constructs information, mutual understanding and its recursive network. Between humans and society, between consciousness and communication there is in fact a fundamental dependency and causal relationship, but at the same time they both operate autonomously and in different (psychic versus social) dimensions of reality.

If we inquire into the conditions of how others can be successfully understood and what the common basis is for taking action and changing reality, then it becomes apparent that, depending on the situation, we resort to common orientation schemes and stocks of knowledge. On the one hand the sign and symbol systems, the rules of human coexistence and the communicative forms of the cooperation, mutual support and conflict are culturally and historically given. They are taught, learned and then shape our plans, our expectations of others and the possibility of understanding others and expressing ourselves. On the other hand the institutionalised uses of signs, social communication forms, socio-cultural structures and situational rules are not only confirmed and maintained, but also continually changed, extended and optimised. In short, communication changes communication. This means (and this validates its regulative and normative claims) that sustainability communication also changes communication and so society.

Through the joint stocks of signs, language(s), values and norms that are produced and reproduced in communication and transmitted through it, social order is built up. The more successful the communicative understanding the more stabile the social order – and vice versa. Nevertheless, successful understanding is not the same as consensus and consensus is not the primary goal or the condition of and for communication. Dissensus is also particularly important for the continuation of communication. Ultimately each communicative act doubles the world and reality towards a yes/no form. "Every communication invites protest. As soon as something specific is offered for acceptance, one can also negate it. The system is not structurally bound to acceptance, not even to a preference for acceptance. Linguistically, the negation of every act of communication is possible and can be understood. It can be anticipated and circumvented by avoiding corresponding communication (…)" (Luhmann 1995: 173).

For sustainability communication both this autonomy from psychic systems and from human intentions as well as the social momentum of orientation and value schemes, consensus/ dissensus and recursive communication sequences is revealing. Sustainability discourse is relatively independent from its many actors who are saying or proposing something. In fact these are interchangeable. It is more decisive *what* and *how* communication takes place. Each act of communication refers to prior acts of communication (accepting or rejecting them) and prestructures at the same time future acts of communication. No longer can everything be said. Expectations arise. This factual-temporal bonding is created by the distinction between theme and contribution (Luhmann 1995).

The social (as well as the non-social) environment enters into communication through themes, which reduce the complexity of the environment to something more specified. In the factual dimension an example would be how the marketing of organic food is concerned with this one particular theme, and nothing else. Communication relationships are ordered by themes, which are, or can be, referred to by various contributions by individuals to communication; and contributions in turn confirm or change themes. In a social perspective themes regulate who can make a contribution, and who is allowed to. And finally the temporal dimension forces a one-by-one *sequentialization* of the contributions to themes. This temporal order allows for continual stream of new references to be made, as well as for a remembrance of past acts of communication and their corresponding system histories. Themes thus take on a memory function. In the 1980s 'sustainable development' began its career as a political semantic and ecological term, and has since served as a reference point for countless discussions, studies and structural changes.

"Thus themes serve as factual/ temporal/ social structures within the communication process, and they function as generalizations insofar as they do not restrict which contributions can be made at what time, in which sequence, and by whom" (Luhmann 1995: 157).

Contributions are themselves re-specifications of themes. As concrete acts of communication they show how themes are interpreted, which information triggers their introduction and who they are relevant for. The political demand that ecological and social aspects be taken into consideration in every situation is just as much a re-specification of the sustainability theme as for example communicating that plane travel should be avoided or publishing an academic text on modern environmental ethics.

A general definition is thus that sustainability communication is a global social process (and one that is accompanied by the mass media) that consists of the recursive order of contributions and arguments to the theme of a better ecological, economic and social life. There are however a number of goals of sustainability communication that are similarly general. Ideally they should be pursued simultaneously (Lass and Reusswig 2001):

- Popularisation goals: the concepts and plans of sustainable development should (not least through mass media support and diffusion) be made known to the general public and offer concrete orientation for action.
- Innovation and alliance goals: Decisive social and technological innovations should be initiated. This would involve a variety of social actors working together and building strategic networks, for example among political parties, business enterprises and NGOs.
- Information and educational goals: Fundamental contents and aspects regarding sustainability should be firmly implemented in the educational system. This would allow children to learn and develop reflexive competence early in life.
- Research goals: Sustainability should become a central research topic in an interdisciplinary scientific discourse with its own perspectives and applications, especially for economic and political actors.

The Nature of Sustainability Discourse

Very few natural hazards and environmental risks are directly experienced by an individual in everyday life. Instead they must first be disseminated by (mass medial) communication. The mass media make the unknown known to the unknown. In the introduction it was pointed out that it is not until sustainability communication reports on human need, wasted resources, potential ecological-economic crises or the lack of rules governing intra- and intergenerational need that these become socially relevant, a social resonance is created and (ideally) remedial action is taken.

Luhmann, in all sociological seriousness, states "that the oil reserves are declining, the rivers are becoming too warm, the forests are dying, the heavens are darkening and the oceans are being polluted. This may be the case, or it may not be the case, but as a physical, chemical or biological fact it will not create any social resonance until it is communicated. Fish may die or human beings; swimming in lakes and rivers may cause illnesses; no more oil may come from the pumps; and average temperatures may rise or fall, but as long as this is not communicated it does not have any effect on society" (1986: 62f.).

Communication and media technology are thus the necessary conditions of sustainability discourse and its social resonance, but this is not to say anything about its typical form and inner structure. In the following some of the characteristics found in sustainability discourse will be discussed and at the same time an analytic framework for its study will be created.

Reflexivity

News about environmental problems or unjust living conditions and research about the destruction of nature and attendant risks to humans have led to public and scientific reaction and reflection, which in turn observes these observations, makes these phenomena and their interrelationships themselves a theme and searches for ways to understand, explain and cope with them. The traditional self-understanding of mankind's currently successful domination of nature and of the evolution of technology is critically examined – and is introduced from society back into society. With this self-referentiality, environmental analysis and sustainability issues become an analysis of society as well as a critique of modern social order (Brand et al. 1997: 37). A further effect of reflexivity is communication about sustainability communication. Sustainability discourse does not just discuss the environment and a better life, but also, and repeatedly, it discusses itself.

Sustainability as an Intrinsic Social Value

Each value is and means a *certain* preference with *universal* validity. Something ought to be, something else ought not; this ranking is fundamentally positive and has a desirable connotation. It stands to reason that we have a preference for freedom, justice, peace, health, conservation etc. and it seems obvious that we have attitudes or make assumptions in favour of them. At the same time values have universal or general validity because they remain, whatever their actual ineffectiveness or non-inclusion, something positive and are (or can be) something that we expect or demand. Their function consists of an action or situation orientation that is neither questioned nor calls for reasons to be provided – this is rarely explicit, much more likely *per implicationem*. "Values remain, in other words, relevant through their allusive nature and that is the source of their infallibility. (…) Values are thus persuasive then because in communication there is a lack of objections; not because one could give reasons for them. (…) Values are the medium for the commonly held assumptions that limit what can be said and what can be wanted, without determining what should be done" (Luhmann 1997: 343). Values compete, however, with each other and depend on particular needs, situations and decisions. That is why they must be dynamically balanced and their application must remain open, i.e. at a given point in time environmental protection instead of freedom, at another welfare instead of intergenerational justice.

Sustainability discourse labours to establish sustainability itself as an intrinsic social value and to gain acceptance for other short-term goals, e.g. securing human survival, inter- and intragenerational justice, maintaining social production potential. On the other hand its value dimensions do not enjoy – everywhere, all the time and without limit – priority over social structures, cultural habits, individual intentions and other values.

The communicated alternatives – of a better life, of anticipatory management, of a just distribution of goods and resources, of a more responsible caring for nature and mastery over nature etc. – are counter-productive when they are connected with an implicit assumption that all too quickly limits or discredits other perspectives and communication contributions, namely that alternatives are always better than what is and what has come before. In addition, sustainability discourse is also labouring to create common perceptions of problems and commitment in the first place, while at the same time there are "a variety of actors struggling with each other to have their own specific definition of sustainability, together with the resulting strategic recommendations, accepted. Behind these disputes are assumptions about different images of the world and nature, different concepts of society, different interests and value preferences" (Brand 2000: 2).

Tendency to Normalisation

The widespread recognition of sustainable development is leading to a normalisation of the concept. The time of ideologically laden struggles is over; objectives are still without doubt being controversially discussed but in general this is being done in a pragmatic fashion. To a great extent this is due to a de-moralisation of environmental issues. This normalisation, de-moralisation and institutionalisation has brought sustainability discourse into a paradoxical situation. The more people talk about and demand sustainability, the less it is able to draw attention to itself or create pressure for change, whether for individual consumers or for key political and economic actors.

Medialisation

Sustainability discourse attempts to resolve the normalisation paradox by linking it with the mass media. It is after all the function of the mass media to generate receptive attention, to inform society, to provide an integrative construction of reality so that there is a reference to common – or at least those assumed to be common – themes, values and knowledge. Through moralising (good vs. bad), the mass media also serves to alarm society. In the mirror of the mass media, society encounters, among other things, its structural problems, is confronted with its catastrophes, ecological risks and, in an extraordinary variety, scandals. What this means for sustainable development is that there are – together with the mass media – two possibilities of educating, warning and improving the public. Either ecological (that is, sustainability) communication itself already implies an attention factor (environmental catastrophe, new data, high profile demonstrations etc.), which are predestined for media coverage and trigger alarm, or sustainability discourse must adapt to the logic of the mass media, must accept journalistic support and medialise itself, so that its communication contributions and visions are broadcast, become known and have consequences. Without effective medialisation there is no popularisation.

Conclusions

The political, moral and scientific discussions centring on sustainable development have not gone unnoticed in the economic field and have triggered a number of reactions in business enterprises and associations. In the course of establishing sustainable objectives, the path between protest movements, NGOs and economic actors has changed in a number of ways, from ignorance to resonance, from confrontation to cooperation. Many economic and other organizations have since taken up sustainability issues, discussed them internally and structurally implemented them in a number of different ways. The genuine communication form of organisations is the *decision* (Luhmann 2000). Within the context of their other commitments and themes, organisations have been able and are able to take sustainability into account in its economic, ecological and/or social dimension (or not!). This means that future decisions are bound by this decision and are thus restricted. "Decision-making programmes define the conditions responsible for the accuracy of decisions" (Luhmann 2000: 257).

Sustainability would then make for a superior decision programme that sets criteria for the evaluation of future projects and organisational objectives. To formulate this more precisely, sustainability functions as an output-oriented goal programme. When sustainability is the goal of what is in principle open-ended planning for the future, then the choice of the possible means (e.g. corporate action) is limited. At the same time by setting such goals the company legitimises its decisions and actions – regardless of whether goals are abandoned or there are unintended consequences or other social values it might pursue (see for example Senge 1999).

Organisations are also necessary in a second case, namely if sustainability is to be addressable, specifically attributable and claimable. When organisations have implemented sustainable development programs they publicise this under their own name, while others as an organisational addressee make demands for sustainability and still other organisations set up guidelines and serve as an addressee for queries or requests for support. Without organised communication, themes cannot be kept visible over the long term, nor specifically attributed, nor disseminated from a specific address.

Sustainability communication is and remains a difficult issue of drawing distinctions and creating resonance. On the one hand there are calls for moderation along with the diagnosis that the environment is unable to fulfil all the demands society evokes and reproduce. And on the other is the highly specialised and functionally differentiated order level of modern society with all of its achievements. At any rate the fact is that no criticism of the risks and consequences of functional differentiation can simply take one side without taking the other into account. "The criticism of functional differentiation remains (…) a moral criticism that cannot account for and cannot determine what otherwise could evolve. That much could be made better is undeniable (…) the apotheosis of one's own morality and the rather unconventional stylistic devices of one's own demeanour might suggest that one should be prepared to revise the assessment. But that will happen anyway and in any case in society and not against it. The secret to those who call themselves alternative is that they do not have any alternatives to offer others. They have to hide this from themselves and others" (Luhmann 1987: 173; see also Rasch 2000).

In the middle of this process of functional differentiation, sustainability communication goes on – precisely because it can handle dissensus. The challenge remains however for communication theory, as well as sociological research, to reflect on ecology in general and sustainability in particular. One of its most important tasks is to continually examine and revise its terminology and theoretical tools and to improve them analytically, so that justice can be done to the complexity of the subject matter – by all means in a fashion that is both critical and enlightening.

References

Brand, K. W. (2000). *Kommunikation über nachhaltige Entwicklung, oder: Warum sich das Leitbild der Nachhaltigkeit so schlecht popularisieren lässt.* Retrieved July 30, 2010, from www.sowionlinejournal.de/nachhaltigkeit/brand.htm.
Brand, K. W., Eder, K., & Poferl, A. (1997). *Ökologische Kommunikation in Deutschland.* Opladen: Leske+Budrich.
Krallmann, D., & Ziemann, A. (2001). *Grundkurs Kommunikationswissenschaft.* München: Fink.
Lass, W., & Reusswig, F. (2001). Für eine Politik der differentiellen Kommunikation – Nachhaltige Entwicklung als Problem gesellschaftlicher Kommunikationsprozesse und –verhältnisse. In A. Fischer & G. Hahn (Eds.), *Vom schwierigen Vergnügen einer Kommunikation über die Idee der Nachhaltigkeit* (pp. 150–174). Frankfurt am Main: VAS.
Luhmann, N. (1986). *Ökologische Kommunikation. Kann die moderne Gesellschaft sich auf ökologische Gefährdungen einstellen?* Opladen: Westdt. Verlag.
Luhmann, N. (1987). Tautologie und Paradoxie in den Selbstbeschreibungen der modernen Gesellschaft. *Zeitschrift für Soziologie, 16*(3), 161–174.
Luhmann, N. (1995). *Social systems.* Stanford: Stanford University Press.
Luhmann, N. (1997). *Die Gesellschaft der Gesellschaft.* Frankfurt am Main: Suhrkamp.
Luhmann, N. (2000). *Organisation und Entscheidung.* Opladen/Wiesbaden: Westdt. Verlag.
Rasch, W. (2000). *Niklas Luhmann's modernity: The paradoxes of differentiation* (Cultural memory of the present). Stanford: Stanford University Press.
Senge, P. M. (1999). *The dance of change: the challenges of sustaining momentum in learning organizations.* New York: Currency/Doubleday.

Chapter 9
Communicating Education for Sustainable Development

Inka Bormann

Abstract There are a number of reasons for communicating education for sustainable development (ESD). One is to use external events to demonstrate the syndrome of unsustainable development in educational contexts; another is to analyse scientific debates on the concept of ESD, including its legitimacy and function and on the political background of the concept of sustainable development. In this context a number of different concepts of ESD are discussed.

Keywords Education for sustainable development • Measurability of education • Criteria and indicators • Competences

Reasons for Communicating Education for Sustainable Development

The ecological catastrophe resulting from the damaged Deep Water Horizon drilling platform in the Gulf of Mexico (beginning April 2010) can be related to lifestyles and consumption patterns in industrialized countries and shows the mutual interpenetration of a number of elements of unsustainable development.

(i) Recognizing this, communicating it and critically reflecting on it can be a reason for ESD – in which for example attention is drawn to the links between this issue and ecological, social and economic problem fields (Michelsen 2005). Initiating ESD programmes using such issues based on the assumption that problems to which the solution is at first beyond an individual's ability to influence are important and legitimate subjects for education. This is in line with Agenda 21 (1992), which served as the starting point for developing the concept of ESD.

I. Bormann (✉)
Phillips University of Marburg, Marburg, Germany
e-mail: inka.bormann@staff.uni-marburg.de

In fact Chapter 36 of Part IV of Agenda 21 is entitled 'Promoting Education, Public Awareness And Training' and calls for a re-orientation of education so that it can better contribute to increasing public awareness of sustainable development. Even though this goal of re-orienting education has been, and still is, the subject of innumerable publications, projects and campaigns in a great variety of educational areas, it is not uncritically shared by all scientists. There is no generally shared definition of the regulative idea of sustainable development, or of the associated concept of ESD, which would help further agreement on goals, tasks, methods etc. nor is it even self-evident that education should be in the service of political objectives, as it would effectively functionalise ESD.

(ii) In view of the at times overly emphatic reference to the concept of sustainable development, Vare and Scott (2007) identify two different forms and objects of ESD, one of which is more and the other less politically ambitioned. The predominant form of practicing ESD is the "promotion of informed, skilled behaviours and ways of thinking, useful in the short term" (ibid, 191). They call this ESD 1, which as "learning *for* sustainable development", is preferred especially by political decision-makers (ibid, 193). ESD 2 is a critical version of ESD 1 and is about "building capacity to think critically about what experts say and to test ideas, exploring the dilemmas and contradictions inherent in sustainable living" (ibid.). Vare and Scott identify this version as best corresponding to the 'real' goal of education as a permanent, open-ended learning process and describe it as "learning *as* sustainable development" (ibid. 194). Vare and Scott thus address one of the main problems of ESD, which according to Bonnett consists of "the notion of SD as a statement of policy" (Bonnett 1999; also Sauvé 1996). This notion leads in turn to – as yet unsolved – semantic, ethical and not least epistemological problems. A brief discussion of this issue from an educational theory perspective as to whether and to what extent ESD can be a legitimate object for educational processes can be found in the third section of this paper. At this point it should be noted that the difference in contextualising ESD is a further reason for communicating scientifically about the concept of education.

(iii) This draws attention to a further aspect. Not only is there dissensus concerning the legitmation of ESD but different approaches can also be identified regarding what exactly is meant by sustainable development and education for sustainable development (IRE 2010). There is a general understanding today regarding the goals of ESD. It is to enable individuals to change their actions so that future generations have a chance to lead a good life (see for example the Bonn Declaration 2009). However, what this exactly means, how this goal can be achieved and how the success of actions can be evaluated are all questions that have widely different answers, as can be seen in the identification of an ESD 1 and an ESD 2. Due to the unequal division of social, ecological and economic resources worldwide, there are not only different claims made on education in general but also differences in the understandings and expectations attached to the functions of ESD. While in Western industrialized countries the issue is more about an 'alphabetisation' in non-sustainable development, in Southern countries it is often about access to basic school education – about the opportunity to be able to read, write and do basic mathematics (EFA – Education for All). And in a report for United Nations Economic Conference for Europe (UNECE) Wals and Eernstman find for the European region that "there is a continuing debate on the meaning of ESD; it is proving difficult to distil the concept in a clear-cut definition, as its interpretation largely depends on the context and the user, and is dynamic in space and time. The only steady characteristic of an ESD process seems to be that it has no universal definition and/or operationalization" (UNECE 2007: §48).

(iv) This situation poses a particular challenge regarding the aspects discussed below. Generally it can be seen that in the educational policy field over the past decades there has been a change in the kind of information used in educational planning and management. In the wake of international comparative studies of educational performance, there is now more uncertainty about the effectiveness of education. One response has been the development of evidence-based forms of management (see for example Scheerens and Hendriks 2004; also Biesta 2007). This trend has since reached ESD and now competences are measured, quality criteria are developed and indicators are formulated in order to investigate the progress of the implementation and success of ESD (e.g. Tilbury 2009; Bormann 2007, 2008; Raaij 2007). A fundamental problem of ESD may be that due to its complexity, lack of definition and clear operationalisation, there are a number of problems involved in undertaking measurement, indicatorisation and evaluation. According to Wals (2009) these can best be dealt with by communication, because then "locally determined indicators, appropriate languages and multiple literacies (…) as well as far more equitable and dialogical forms of interaction" (ibid, 195) can be realised (see Bormann and Michelsen 2010).

After this brief overview of four reasons for communicating ESD, this chapter will now concentrate on the fourth aspect. The focus will especially be on the currently dominant form of communicated knowledge of ESD, that is on methodologically controlled knowledge as used in various fields of action in decision-making and management processes.

Towards the Communication of the Effectiveness of ESD

Educational policy institutions no longer uncritically assume that interventions lead to their intended consequences. Instead as a part of education monitoring and accountability (Anderson 2005) a variety of instruments should enable the provision of a rational and data-based description of the actual state of the effectiveness or ineffectiveness in and of educational systems as well as whether it is necessary to take any appropriate action.

As late as the 1980s educational policy management largely used input indicators to improve the educational system. The international standard today however revolves around output-oriented, evidence-based management. Educational monitoring involves a number of different objects and levels and extends from operationalisation and measurement of individual competences to criteria-supported observation of the organisational structures of teaching and learning to indicator-based observation of the performance of the whole system (Rode and Michelsen 2008).

Evidence-based methods have since reached the field of education for sustainable development. The discussion is centred on how to best investigate and communicate sustainability development, which competences learners can acquire in the context of ESD (Transfer 21a 2007; Bormann and Haan 2008), which competences teachers should have when 'teaching' ESD (UNECE 2009), which 'sustainable' characteristics educational organisations should possess (Transfer 21b 2007; Breiting et al. 2005) and how ESD can be anchored in the educational system (Wals 2010; Tilbury 2009; UNECE 2007; for sustainability indicators see Davidson 2010).

Measurability of Education

Reality exists without systematic empirical observation. And it resists any definite, quasi-ontological meaning (Meyer-Drawe 1999: 329). Thus, in trying to observe 'reality' there is a tension between an *explanandum* – a thing needing to be explained – and its *explanans* – a set of claims that will explain what needs to be explained. In terms of monitoring education, there is a tension between the phenomenon of education, which takes place every day, formally, non-formally and informally, and its social constitution, a more or less commonly shared understanding of what education is, e.g. in the form of definitions, descriptions, models or indicators.

The term education is used both normatively (Tenorth 1997) and descriptively and thus definitions or descriptions of the concept include a number of different connotations, such as

- Input: this indicates an institutional orientation – how much money is invested into the education system? Education appears as a 'regulated task'.
- Processes (concerning both educating and being educated): this is a time-oriented understanding – how is education organised, how much time is spent on particular subjects etc.? Education appears here as 'organised appropriation'.
- Output and outcome: this is a results-oriented understanding – what abilities, skills, knowledge and competences have been acquired or achieved? Education appears as a 'product'.

Depending on the emphasis given to these diverse connotations, the functions attributed to education might range from something that refers to the individual, to an anthropological concept or to an understanding that considers education to be a task of educational institutions and organisations.

The *explanandum* education is thus indisputably a complex phenomenon, which has a variety of persons or groups responsible for it as well as diverse persons or groups to which it is addressed. The term education does not seem to classify the phenomenon in a distinctive, proper manner. Rather it mirrors the need for reflection on the content of education, which needs to be accounted for, as well as deliberation on the adequacy of the available resources and tools to measure it.

The heterogeneity of meanings and functions also shows that the term education comprises different knowledges. It is a truism that it is impossible to discern any single detail of a complex *explanandum*. A condition for its analysis is that it be reduced, or operationalised. This is accompanied by (over)simplifications, as well as by a selective accentuation of aspects considered to be, or negotiated as, important.[1] The operationalistic reduction of education in conjunction with the monitoring of educational systems can be done by means of indicators. These would allow the acquisition of knowledge about selected aspects of the concept education.

An anthropological perspective studies the social meaning of education and assumes a subconscious and practical knowledge that avoids being easily fixed. It is apparent that from this perspective the recent occurance of the 'evaluative habitus', the measurement of education, will be viewed critically. Measuring, in particular measuring and assessing outputs, is accompanied with reductionism, the reduction of education to performance variables (Radtke 2003). This is also true for ESD.

Towards the Measurement of Education for Sustainable Development

Considering the many different goals and contents of ESD, it would be problematic to make general statements about ESD. In addition there are many didactic possibilities and places of learning, non- and informal learning as well as many addressees of ESD, including children and young people, teachers, teacher trainers, consumers, organisational leaders as well as political and administrative decision-makers. All of these target groups will have a more or less clear idea about the goals of ESD, but they will also have their own goals as well and will be more or less willing and able to actively pursue them. Recent research also shows that competences are situationally upgraded in connection with so-called domain-specific knowledge, that is by means of concrete requirements and specifications. And since the lives of these groups of individuals will without doubt be different it will not be possible to assume that they have the same competences – much less that these competences can be investigated in detail with a 'one-size-fits-all' instrument.

Learning conditions can be regulated to different degrees and as such the spectrum of what could be expected is quite large. While in the formal educational sector for example the attempt could certainly be made to implement contentual aspects of ESD, in the non-formal educational sector this is by definition impossible. To this extent it is extremely difficult to make generalisations about the effectiveness of measures regarding the acquisition of competences or, on the systemic level, about 'the' quality of ESD.

---

[1] Insofar the chosen tools are not nearly just neutral measures or signs of the consequences of an intervention (Frønes 2007: 20).

Competences

A competence is in general considered to be the ability of an individual to successfully deal with the demands he is confronted with in a particular context. More specifically, competences are those cognitive abilities and skills an individual has or can learn in order to solve specific problems together with the related motivational, volitional and social willingness to apply the problem solution in a variety of different situations (Weinert 2001). Such a broad definition of competence is oriented toward the ideal of a comprehensive capacity to act and individual maturity (Jude et al. 2008).

The orientation towards competences directs attention to learning outcomes, or outputs, as can be seen for some time in both public and academic debates. The demand that school pupils have defined competences at a specific point in time of the learning process is not a unique characteristic of education for sustainable development. For a number of years there has been a demand for pupils in schools to be educated so that they can lead a successful life and actively take part in social developments. The key competences thought to be needed can be found in an OECD reference framework (Rychen and Salganik 2003). Sustainability, alongside human rights, equity and social cohesion, is one of the social goals behind the OECD conceptual framework of 'definition and selection of competencies' (DeSeCo).

One of the broad categories of competencies that an individual needs is – in addition to the ability to use tools (including language) interactively and to be able to engage with diverse groups in a globalised world – to act responsibly as an autonomous individual within a broader social context. These three primary categories of competencies in the OECD reference framework are themselves oriented towards the classic dimensions – as discussed in the discourse on competences since the 1970s – of subject knowledge and method, social and personal competence.

With these provisions the DeSeCo concept is a suitable point of reference specifically for competency models designed for education for sustainable development. Thus, there is frequent reference to it when attempting to reach agreement on which competencies are necessary (see the contributions to Bormann and de Haan 2008). It does not however provide a unitary concept for developing these competencies for ESD. Instead there is a wide variety of concepts by means of which the competencies of ESD could be determined and modelled (ibid. 8), each showing great differences and different motivations regarding its contentual requirements.

Competency models specify the contents and goals of educational programmes and to this extent are points of reference for developing teaching and learning processes. There are quite different thoughts on what competencies are, how they can be derived, justified, described and measured. For example the GRF Priority Programme 'Competence Models' has focused on cognitive dispositions (Klieme and Leutner 2006: 4; Hartig et al. 2008). Other concepts add the affective and motivational dimension to the cognitive, with reference frequently being made to Weinert's (2001) definition of competence.

Furthermore, concepts of competence differ along basal directions. Schecker and Parchmann (2003) propose distinguishing between *descriptive and normative* models as well between competence *structural* models and competence *development* models. In a competence structure model individually desirable components are defined in relationship to a primary contentual goal. They give information about the requirements necessary for learners to be able to cope with tasks and problems in a specific domain or requirement area. Competence *development* models go beyond the structural models to the extent that, on the basis of such conditions as learning environment and experience as well as a number of contentual requirements in a knowledge domain, they take individual components of a more complex competence and order them in temporal or developmental hierarchy. Although there are such implicit development models, e.g. in the form of curricula, it remains to be investigated which cognitive requirements are necessary in order for declarative knowledge to be transformed into procedural knowledge.

There are no concepts that include the *development* of sub-competences in terms of *Gestaltungskompetenz*[2] (de Haan 2006, de Haan et al. 2009) in ESD. In ESD competence structural models both these points can be identified. There are models that largely concentrate on the cognitive dimensions of competences, for example, the model developed at the Leibniz Institute for Science and Mathematics Education with its components of 'understand/know-evaluate-act' (Lauströer and Rost 2008). A further study that is essentially oriented towards the knowledge component is the study 'Green at Fifteen?' (OECD 2009). The concept of *Gestaltungskompetenz* is a holistic competence concept (Wals 2010: 149) that also includes the social and affective dimensions.

The concept of *Gestaltungskompetenz* is a normative competence structural model based on the OECD reference framework. It specifies the functions of ESD and now includes 12 related sub-competences, which in turn can be classified in the three primary competence categories of the OECD framework (Table 9.1).

The attention that competences have received for some time now shows an increasing orientation to research findings into the output of learning and educational processes. But the investigation into whether and to what degree the proposed competences can be acquired is still at the beginning – and has proved to be quite difficult. This is also due to the variety of different approaches used, e.g. descriptive or normative; competence structural or developmental models. Especially for everyday topics – and this is exactly what education for sustainable development is about – it is a particular challenge to formulate empirically valid competence levels. For these domains, according to Klieme (2004), "there may be no levels that can be clearly demarcated and put on a scale from 'low' to 'high', but rather different patterns or

---

[2] *Gestaltungskompetenz* describes the ability "to apply knowledge of sustainable development and recognise problems of non-sustainable development. That means drawing consequences from analyses of the present and future scenarios on environmental, economic and social developments in their interdependence to take decisions and understand them before implementing them as individuals, in the community and politically in a way to promote sustainable development processes" (Transfer 21a: 12).

Table 9.1 Classification 'Gestaltungskompetenz' sub-competences (de Haan et al. 2009)

Competence categories of OECD	Sub-competences of 'Gestaltungskompetenz'
Use tools and media interactively	T.1 *Competence for perspective-taking:* Be open-minded and create knowledge from new perspectives
	T.2 *Competence for anticipation:* undertake forward-looking analysis and evaluate developments
	T.3 *Competence for interdisciplinary knowledge acquisition:* acquire interdisciplinary knowledge and act on it
	T.4 *Competence for dealing with incomplete and overly complicated information:* recognize risks, dangers and uncertainties and be able to evaluate them
Interact in homogenous groups	G.1 *Competence for cooperation:* be able to plan together with others and take action
	G.2 *Competence to deal with individual decision-making dilemmas:* account for conflicts in goals when reflecting on action strategies
	G.3 *Competence for participation:* be able to take part in collective decision-making processes
	G.4 *Competence for motivation:* be able to motivate one's self and others to take action
Act autonomously	E.1 *Competence for reflecting on goals:* be able to reflect on one's own goals and those of others
	E.2 *Competence for moral action:* be able to use ideas of justice as a basis for making decisions and taking action
	E.3 *Competence for independent action:* be able to independently plan and act
	E.4 *Competence for supporting others:* be able to show empathy towards others

types" (ibid, 13). And for the investigation of competences, Jude et al. (2008) make the fundamental point "that the high expectations for a competence diagnostic (…) are confronted with measurement methodologies (…) that are still unsatisfactory" (ibid, 7).

Criteria and Indicators

Competence models have the goal of providing statements about individual learning results and abilities resulting from ESD or normative orientation about structuring what should be learnt and how. Criteria and indicators on the other hand aim at observing progress in development at the level of the organisation or the whole system and when applicable enable comparisons or inform decisions to be made.

Criteria and indicators were first used in economic and social reporting. Indicators are common instruments for evaluating guidelines, measures and programmes. They serve the preparation of political and administrative decision-making (Ben-Arieh and Frønes 2007). However, "there is no single, generally applied definition of 'indicator'" (de Vries 2001: 319). In fact, there is a controversy about defining the term 'progress' as an indicator (de Vries 2001; Goulet 1992). In extremely simplistic and general terms, an indicator or a system of indicators designates something that yields conclusions about another thing that cannot be directly observed. In contrast to this very broad understanding of indicators (e.g. Fitz-Gibbons and Tymms 2002), there is another, more narrow definition, which requires that indicators be validated in order to obtain reliable statements about real or simulated developments and on this basis to take political measures (e.g. Kaplan and Elliott 1997).

At the latest since the 1980s when findings from research on how scientific knowledge is put to use in practice, there have been critics who would like to 'demystify' such a position with the empirically based argument that scientific knowledge gained through standardised methodologies is much less able to be integrated into political or institutional actions than was once assumed (Beck and Bonß 1989).

But for a number of years a trend toward 'indicatorisation' in the area of ESD has been observed. In the late 1990s indicators began to be developed for use in ESD and have since been developed for a number of different contexts, addressees and purposes.

While these criteria are intended for use on the meso-level, i.e. in educational organisations, the indicator set of UNECE is oriented towards governments. Its 48 qualitative and quantitative sub-indicators operationalise the six goals formulated in the UNECE strategy together with the UNESCO (UNECE 2005; UNESCO 2005). This strategy is meant to encourage the countries of the UNECE region to integrate ESD in their educational systems. The indicators in contrast are meant to support observation of the progress made in implementing them (UNECE 2005: §6; Tilbury 2009; Pigozzi 2010).

The ESD indicators currently being used differ in a number of ways from those developed for and used in general educational systems monitoring. This is especially due to the fact that ESD is a modernisation concept that is cross-cutting and does not yet have a completely institutionalised identity and as such can still be considered something new.

Communication About ESD

There are certainly sceptical voices about the legitimacy of ESD. One criticism is that ESD lacks a grounding in educational theory, and is, in contrast, rather politically driven or instrumentalised. Although elaborated long before ESD became an issue in scientific research, remembering Klafki's concept of critical-constructive didactics still seems a helpful orientation. In the 1960s Klafki undertook to reclassify general education by identifying topics, or so-called 'key problems of the epoch', that should play a central role in modern general education. These topics or key problems range from the question of peace to the consequences of technology, equal rights and democratisation to the environment. Dimensions and problems of sustainable and unsustainable development can certainly be integrated into this complex.

A confrontation and discussion on these inter- and transdisciplinary topics should not lead to 'only' a material education, i.e. the acquisition of canonised subject matter, or 'only' a formal education, i.e. the development and application of competences. When dealing with such comprehensive subjects, static learning content should not be in the foreground nor according to Klafki should the abilities and skills trained in discussion become an end in themselves. The goal is a categorial education, i.e. education at the same time individual and social, problem-based, problem-developing and problem-solving. For this to happen, Klafki proposed that for each topic made an object of educational processes plausible answers be given to the following five questions:

1. To what extent is the object exemplary for a general problem?
2. How important is the topic for the life of learners today?
3. How important will the topic be for the life of learners in the future?
4. How can the content of the topic be structured by means of questions 1 and 2?
5. How can the topic be clearly presented so that it is accessible to learners? (Klafki 2007)

The problems of sustainable and unsustainable development can clearly be given a logical theoretical framework and the danger of ESD being political misused can be mitigated. At the same time it can be argued that education – including ESD – is more than what can be measured by indicators as learning outcomes. A further criticism is directed, in the context of the competence debate, at the functionalisation of ESD. This criticism implies that ESD is a 'means to an end', i.e. is being used for the solution of political problems in education. From the perspective of Klafki's concept of education it also becomes clear that specifying *Gestaltungskompetenz* as a measurable outcome of ESD can be an orientation for the organisation and the evaluation of educational processes, but to restrict the objective of ESD to the teaching of *Gestaltungskompetenz* or to appraise its value in regard to the declarative knowledge gained is too limited. In spite of the functionality in the concept of ESD and in the orientation to competences and indicators, ESD should be primarily understood as a comprehensive educational concept that is directed at the self-formation of a responsible and active personality. As such ESD is a much broader concept than what is taught as a competence, and, it is much more than numbers occuring in indicatorisation exercises.

Thus, communication on ESD seems a never-ending or rather: long-lasting issue, after all. An end to communication about the concept of ESD, its implementation in various fields of action, its 'usability' and legitimacy, is nevertheless not in sight. That would also be regrettable as it would also mean an end to its critical-constructive development. In future it will probably be more a question of how to develop ESD research and practice and communicate it so that the object ESD is released from its niche existence and instead is recognised as putting forward its own contributions to the solution of urgent social problems as well as addressing recent scientific research questions.

References

Agenda 21 (1992). *Agenda 21, New York: United Nations*. Retrieved July 30, 2010, from www.un.org/esa/sustdev/documents/agenda21/english/Agenda21.pdf.

Anderson, J. A. (2005). *Accountability in education*. Paris: UNESCO.

Beck, U. & Bonß, W. (Eds.). (1989). *Weder Sozialtechnologie noch Aufklärung? Analysen zur Verwendung sozialwissenschaftlichen Wissens*. Frankfurt: Suhrkamp.
Ben-Arieh, A., & Frønes, I. (2007). Indicators of children's well-being – concepts, indices and usage. *Social Indicator Research, 80*, 1–4.
Biesta, G. (2007). Why 'what works' won't work: evidence-based practice and the democratic deficit in educational research. *Educational Theory, 57*(1), 1–22.
Bonnett, M. (1999). Education for sustainable development: a coherent philosophy for environmental education*? Cambridge Journal of Education, 29*(3), 313–324.
Bormann, I. (2007). Criteria and indicators as negotiated knowledge and the challenge of its transfer. *Educational Research for Practice and Policy, 6*(1), 1–14.
Bormann, I. (2008). Fortschrittsmonitoring mittels Indikatoren - ein Beispiel. In W. Böttcher, W. Bos, H. Döbert & H. G. Holtappels, (Eds.), *Bildungsmonitoring und Bildungscontrolling in nationaler und internationaler Perspektive* (pp. 47-58). Münster: Waxmann.
Bormann, I., & Haan, G. (Eds.). (2008). *Kompetenzen der Bildung für nachhaltige Entwicklung* (Operationalisierung, Messung, Rahmenbedingungen, Befunde). Wiesbaden: VS.
Bormann, I., & Michelsen, G. (2010). The collaborative production of meaningful measure(ment) s. In: *EERJ* 9 (4).
Breiting, S., Mayer, M., & Mogensen, F. (2005). *Quality criteria for ESD schools*. Wien: Bundesministerium für Bildung, Wissenschaft und Kultur.
Davidson, K. M. (2010). Reporting systems for sustainability: what are they measuring? *Social Indicators Research* 86, online first: www.springerlink.com/content/d68n31n7385r8rt1/?p=1d 1ac513dfd3474c99a79270a9de25bb&pi=12.
de Haan, G. (2006). The BLK '21' programme in Germany: a 'Gestaltungskompetenz'-based model for education for sustainable development. *Environmental Education Research, 12*(1), 19–32.
de Haan, G., Kamp, G., Lerch, A., Martignon, L., Müller-Christ, G., & Nutzinger, H. G. (2009). *Nachhaltigkeit und Gerechtigkeit*. Berlin: Springer.
de Vries, M. (2001). Meaningful measures: indicators on progress, progress on indicators. *International Statistical Review, 69*(2), 313–331.
Declaration, B. (2009). *Journal of Education for Sustainable Development, 3*(2), 249–255.
Fitz-Gibbons, C. T., & Tymms, P. (2002). Technical and ethical issues in indicator systems: doing things right and doing wrong things. *Education Policy Analysis Review* 10 (6). Retrieved July 30, 2010, from http://epaa.asu.edu/epaa/v10n6.
Frønes, I. (2007). Theorizing indicators. On indicators, signs and trends. *Social Indicators Research, 83*, 5–23.
Giddens, A. (1997). *The constitution of society*. Frankfurt: Campus.
Goulet, D. (1992). Development indicators: a research problem, a policy problem. *Journal of Socio-Economics, 21*(3), 245–261.
Hartig, J., Klieme, E., & Leutner, D. (2008). *Assessment of competencies in educational contexts: state of the art and future prospects*. Göttingen: Hogrefe & Huber.
IRE International Review of Education (2010). Special Issue , Education for sustainable development, 56 (2–3). In Bormann, I., Haan, G. de & Leicht, A. (Eds.).
Jude, N., Hartig, J., & Klieme, E. (2008). *Kompetenzerfassung in pädagogischen Handlungsfeldern. Theorien, Konzepte und Methoden*. Berlin: BMBF. Retrieved July 30, 2010, from www.bmbf. de/pub/bildungsforschung_band_sechsundzwanzig.pdf.
Kaplan, D., & Elliott, P. R. (1997). *Journal of Educational and Behavioral Statistics, 22*(3), 323–347.
Klafki, W. (2007, 6.A. [1963]). *Neue Studien zur Bildungstheorie und Didaktik*. Weinheim: Beltz.
Klieme, E. (2004). Was sind Kompetenzen und wie lassen sie sich messen? *Pädagogik, 56*(6), 10–13.
Klieme, E. & Leutner, D. (2006). *Kompetenzmodelle zur Erfassung individueller Lernergebnisse und zur Bilanzierung von Bildungsprozessen*. Antrag an die DFG zur Errichtung eines Schwerpunktprogramms.
Lauströer, A., & Rost, J. (2008). Operationalisierung und Messung von Bewertungskompetenz. In I. Bormann & G. de Haan (Eds.), *Kompetenzen der Bildung für nachhaltige Entwicklung. Operationalisierung, Messung, Rahmenbedingungen, Befunde* (pp. 89–103). Wiesbaden: VS.
Meyer-Drawe, K. (1999). Herausforderung durch die Dinge. Das Andere im Bildungsprozess. *Zeitschrift für Pädagogik, 45*(3), 329–336.

Michelsen, G. (2005). Nachhaltigkeitskommunikation. Verständnis – Entwicklung – Perspektiven. In G. Michelsen & J. Godemann (Eds.), *Handbuch Nachhaltigkeitskommunikation. Grundlagen und Praxis* (pp. 25–42). München: Oekom.

OECD. (2009). *Green at fifteen? How 15-year-olds perform in environmental science and geoscience in PISA 2006.* Paris: OECD.

Pigozzi, M. J. (2010). Implementing the UN decade of education for sustainable development (DESD). Achievements, open questions and strategies for the way forward. *International Review of Education* 56. Retrieved July 30, 2010, from www.springerlink.com/content/b50u1 34666036522/?p=bc288a34a7cc467fb98cee6c05ffd1e2&pi=4.

Raaij, R. van (2007). Indicators for education for sustainable development. *Education for Sustainable Development* 1(1). Retrieved July 30, 2010, from www.bne-portal.de/coremedia/generator/pm/en/Issue__001/01__Contributions/Raaij_3A_20Indicators_20for_20Education_20for_20Sustainable_20Development.html.

Radtke, F.-O. (2003). Die Erziehungswissenschaft der OECD – Aussichten auf die neue Performanzkultur? *Zeitschrift für Erziehungswissenschaft, 14*(27), 109–136.

Rode, H., & Michelsen, G. (2008). Levels of indicator development for education for sustainable development. *Environmental Education Research, 14*(1), 19–33.

Rychen, D. S., & Salganik, L. H. (Eds.). (2003). *Key competencies for a successful life and well-functioning society.* Göttingen: Hogrefe & Huber.

Sauvé, L. (1996). Environmental education and sustainable development: a further appraisal. *Canadian Journal of Environmental Education, 1,* 7–34.

Schecker, H., & Parchmann, I. (2003). Modellierung naturwissenschaftlicher Kompetenz. *Zeitschrift der Didaktik der Naturwissenschaften, 12,* 45–66.

Scheerens, J., & Hendriks, M. (2004). Benchmarking the quality of education. *European Educational Research Journal, 3*(1), 101–114.

Tenorth, H. (1997). "Bildung" – Thematisierungsformen und Bedeutung in der Erziehungswissenschaft. *Zeitschrift für Pädagogik, 43,* 969–984.

Tilbury, D. (2009). Tracking our progress. A global monitoring and evaluation framework for the UN DESD. *Journal of Education for Sustainable Development, 3*(2), 189–193.

Transfer 21a (2007). *Education for sustainable development at secondary level. Justifications, competences, learning opportunities.* Berlin: Freie Universität Berlin. Retrieved July 30, 2010, from www.transfer-21.de/daten/materialien/Orientierungshilfe/Guide_competences_engl_online.pdf.

Transfer 21b (2007). *Developing quality at "ESD Schools". Quality areas, principles & criteria.* Berlin: Freie Universität Berlin. Retrieved July 30, 2010, from www.transfer-21.de/daten/materialien/Orientierungshilfe/quality_eng_online.pdf.

UNECE (2005). *UNECE strategy for education for sustainable development.* Retrieved July 30, 2010, from www.unece.org/env/documents/2005/cep/ac.13/cep.ac.13.2005.3.rev.1.e. pdf.

UNECE (2007). *Learning from each other: Achievements, challenges and the way forward.* Retrieved July 30, 2010, from www.unece.org/env/documents/2007/ece/ece.belgrade.conf.2007.inf.3.e.pdf.

UNECE (2009). *Proposal for the establishment of an expert group on competences in education for sustainable development.* Terms of Reference of the Expert Group on Competences. Retrieved July 30, 2010, from www.unece.org/env/esd/SC.EGC.htm#background.

UNESCO (2005). *International implementation scheme.* Retrieved July 30, 2010, from unesdoc.unesco.org/images/0014/001486/148654E.pdf.

Vare, P., & Scott, W. A. H. (2007). Learning for a change. Exploring the relationship between education and sustainable development. *Journal of Education for Sustainable Development, 1*(2), 191–198.

Wals, A. E. J. (2009). A Mid-DESD review: key findings and ways forward. *Journal of Education for Sustainable Development, 3*(2), 195–204.

Wals, A. E. J. (2010). Between knowing what is right and knowing that it is wrong to tell others what is right: on relativism, uncertainty and democracy in environmental and sustainability education. *Environmental Education Research, 16*(1), 143–151.

Weinert, F. E. (2001). Concept of competence. A conceptual clarification. In D. S. Rychen & L. H. Salganik (Eds.), *Defining and selecting competencies* (pp. 45–65). Seattle: Hogrefe & Huber.

Chapter 10
Sustainability Communication: A Systemic-Constructivist Perspective

Horst Siebert

Abstract Systems theory and constructivism as background theories are widely discussed. This chapter gives an overview of the core theses, key terms and observer perspectives of this paradigm, with reference to the special features of systemic thinking. The interconnectedness of cognition and emotion as well as the construction of reality through language are important for the sustainability discussion.

Keywords Systems theory • Constructivism • Cognition • Emotion • Constructive epistemology

Niklas Luhmann's Systems Theory

Systems theory and constructivism are common 'background theories' and are used in education, therapy, social work and media communication. One of the best known proponents of systems theory is the sociologist Niklas Luhmann. His core thesis is that the more complex the overall social system, the more important the functional subsystems, such as the health, educational, science systems etc. In order for such a subsystem to operate functionally, the boundaries between system and environment (not in an ecological sense) have to be clear-cut.

Systems are characterized by autopoiesis (self-organization) and self-reference. Both social systems and psychic systems (that is, human consciousness) are self-referential. They generate their own value standards from themselves, that is from their own history and experiences, even though they are continually confronted with external referential demands. But it is because of their self-referential nature that a system (e.g. the school system) can handle such external expectations in a

H. Siebert (✉)
Leibniz Universität Hannover, Germany
e-mail: sarah.koehler@ifbe.uni-hannover.de

relatively independent fashion and make its own contribution to the stability of the system.[1]

Systems are distinguished by meaningful operative distinctions, which can be defined as binary codes (e.g. true/not true for the science system). They are operationally closed, that is they do not have any direct contact with the outer world, but they are structurally coupled with the environment. Subsystems provide services for society. For example, the educational system allows individuals to earn occupational qualifications.

Luhmann's systems theory is a sociological super theory that describes and explains all social areas with the same conceptual tools – structure, function, operation, differentiation, medium, operative distinction etc. Luhmann analyses how society processes complexity and how subsystems must be structured if they are to fulfil their functions. This theory does without such constructs as 'rational behaviour' and without normative settings – which does not at all mean that Luhmann is indifferent to injustice, exploitation or environmental destruction. Luhmann achieves a theoretical shift in perspective that is also inspired by constructivism. We no longer look for the ontological essence of things, but for the epistemic limits to knowledge. Luhmann links systems theory with epistemology with his system-environment model: "It has long been known that the mind has no qualitative and very little quantitative contact with the environment. The whole nervous system simply observes the changing states of its organism and not what happens outside it" (Luhmann 1990: 36ff). The system-theoretical point of this finding of brain physiological research is an apparent paradox: "Only closed systems can understand (…). We can only see because we cannot see (…). The effect of this intervention from systems theory can be described as a de-ontologisation of reality. It does not mean that reality is denied (…) It is the epistemological relevance of an ontological representation of reality that is contested (…) A further consequence is that no system can complete operations outside of its own boundaries" (Luhmann 1990).

At the same time the concept of society is 'deontologised'. Luhmann's most general description of society is that society – including global society – is communication. The individual subsystems are differentiated according to their specific (and changeable) communication code. 'Sustainability' has become the meaningful operational distinction of the 'ecological subsystem'. (Whereas it remains to be determined whether ecology – similar to education – has the characteristics of an independent 'system' or whether the construct 'sustainability' is contributing to the disintegration of the 'ecosystem' by de-differentiation.) At any rate the medium of ecological communication is no longer 'conservation' but 'sustainable development'.

Systemic Thinking

It is good practice to distinguish 'system-theoretical' and 'systemic', even though both are related. Relatively uninfluenced by Luhmann's theoretical structure, Vesta, Capra, Dörner et al. have pleaded for a systemic thinking. Their argumentation – much

---

[1] The following anecdote may help to clarify this. A balloonist is lost when he sees a farmer in a field below. He calls out to him: 'Where am I?' The farmer shouts back, 'In a balloon'.

simplified – is that complex social, political, economic and ecological systems are characterized by reciprocal effects, effect networks (instead of effect chains), feedback couplings, butterfly effects, incalculable side and follow-on effects and chaotic turbulences. A linear, dualising (either-or), calculating, mechanistic thinking (which is suitable for modern technology and industry) does not do justice to such networking. A 'flexible intelligence' is needed, one that does not search for final, irreversible solutions (nuclear power plants are irreversible in that they cannot be 'undone'), but that can 'carefully' handle ambiguous, circular processes, insecurities and uncertainties. Systemic thinking is thinking in contexts. Systemic ecological thinking finds a touchpoint with systemic therapy and systemic pedagogy (Balgo and Werning 2003). For environmental education this means context sensitivity. For example explaining to high school students how their own behaviour can be environmentally destructive is to simultaneously question their social, cultural, biographic and economic interrelationships. Therapy and counselling have been strongly influenced by systemic-constructivist paradigms. Family therapy in particular no longer 'treats' individual family members but views circular interactions within a family system. The therapist foregoes the pretence of understanding those involved better and more deeply than they themselves do (by uncovering the subconscious). Feelings and thoughts can only be perceived by each individual himself. Childhood stories are also no longer a central element of therapy. A key concept and instrument of systemic therapy is observation. How do family members observe (and construct) themselves and each other? As an observer of the second order, the therapist registers how those involved observe each other, while at the same time reflecting on his own role as observer. He also has blind spots and he does not know any 'objective' facts. Simon asks self-critically: "Are not phenomena also constructed through observation or by the observer himself? (…) Doesn't the system 'family therapist' also need to be observed?" (1997: 13). Simon concludes that "The most important premise one has to leave behind is the assumption that one could make any independent, that is 'objective', statements about any patient that are independent from the conditions of observation" (1997: 14).

Constructive Epistemology

Systems theory and constructivism are closely related. As disciplines, systems theory is a social science and constructivism is a part of epistemology. The founders of neuro-biological constructivism, the biologists Maturana and Varela, argue from a systems theoretical perspective. Luhmann also examined constructivism more closely in his later work. The explanation for this fusion has to do with the comprehensive definition of system. Luhmann defines society as a (social) system, but also human consciousness as a (psychic) system and the human organism as a (biological) system. Autopoiesis, self-reference (that is recursivity) and operational closure are concepts also made use of in these systems.

Developments in cognitive and neuro-sciences have 'strengthened' the thesis that our brain is a self-organised, structurally determined system that does not represent the reality outside subjectivity 'truly' but instead constructs realities of its

Fig. 10.1 Constructivist disciplines

Epistemological theory	Brain theory	Biology
Cognitive psychology	Psychology of emotion	Social psychology
Systems theory	Cultural theory	Linguistics

own kind through its senses, as well as cognitively and emotionally. This thesis is also fruitful for pedagogical and communication sciences (Fig. 10.1).

The world is not directly accessible by our knowledge system. Even our visual, acoustic, olfactory and tactile perceptions are translations of chemical and physical stimuli. For example, hearing is a translation of sound waves. Our brain may be 'structurally coupled' with the environment, but 'inputs' from the environment are merely triggers for highly individual and experience-dependent thought processes. The system itself decides what it considers to be relevant input. In this perspective learning is also a self-regulated, emerging construction of reality. Although adults are capable of learning, they are 'unteachable',[2] that is they cannot be directly instructed or raised like little children.

Teaching and learning are structurally separate systems. Learning systems cannot be determined from outside. They can only perceive and process information that can be processed by the cognitive system. To formulate it as a paradox: We do not see what we do not see; we hear only what we hear. "In the psychological and especially in the biological testing of this hypothesis, we have discovered that every brain constructs, that is to a certain degree discovers 'reality' differently" (Lenzen 1999: 155).

In a seminar 20 students will then construct 20 different ideas of the learning content of the seminar, so that each student is learning something different. This means that we cannot 'educate' someone to be moral. Ecological responsibility must be experienced and lived as self-referential.

Cognition and Emotion

In the 1990s constructivism was mainly discussed as a theory of cognition. In recent years – not least because of advances in brain research – it has become clear that our constructs of reality are just as much emotional as cognitive. Roth, a brain researcher,

---

[2] As the saying goes, "You can lead a horse to water but you can't make it drink".

attributes a controlling function in human behaviour to emotions: "Reason and understanding are embedded in the affective and emotional nature of humans. The largely unconscious centre of the limbic system is not only formed much earlier than the conscious cortical centre. It builds a framework within which the others can work (...) Consciousness and understanding can only be turned into action with the 'approval' of the limbic system" (Roth 2001: 451ff.).

This shift of emphasis from the cognitive to the emotional control of action is of central importance for ecological sustainability. It is however a pedagogical dilemma that the term sustainability is abstract and theoretical and rather 'emotionless' and thus – in contrast to for example 'nature' or 'animal welfare' – it is hardly suited for triggering 'sensual' feelings. This could be an explanation for why the term sustainability has remained relatively ineffective in educational practice.

The Swiss emotion psychologist and constructivist Luc Ciompi emphasizes the effectiveness of 'affective communication' and pleads for a reorientation in science and pedagogy. Ciompi draws attention to the action guidance effects of logics of human affect. Basic emotional moods (happiness, fear, sadness) colour our sensory perceptions, thoughts and feelings. This means that "according to the logic of affect, affective and cognitive components join together with the accompanying senso-motoric system to functionally integrated affective-cognitive reference systems or feeling-thinking behaviour programmes" (Ciompi 2003: 62). These behavioural programmes are similar to Piagetian 'schemes'. We thus do not learn merely facts, terms, 'subject matter', but instead programmes that make up our identity, that are emotionally anchored, that allow an action orientation. Action is not only a result of cognitions, but successful actions are at the same time knowledge processes. The concept of feeling-thinking behavioural programmes appears to be promising in promoting sustainable development. Such 'programmes' can include, e.g. water, energy, forest, waste etc.

Language as Construction

Radical constructivism stresses the uniqueness and isolation of the thinking individual. Humans are 'opaque' to each other, so that Luhmann and Roth both note that the normal case for communication is misunderstanding. On the other hand communication is – in spite of all the attendant difficulties in understanding – necessary for survival. Humans are not only viable as social creatures but also live in interpretation communities. Even if each individual observes something else, these observation perspectives and interpretation patterns have a collective basis. Symbolic interactionism had already pointed out that our perceptions and interpretations arise – through socialization – in social contexts; through contact with new reference groups previous images of self and the world are modified.

These findings from social-psychological research are the starting point for social constructivism. Gergen, a well-known proponent of social constructivism, draws attention to the social construction of reality through language by pointing out that

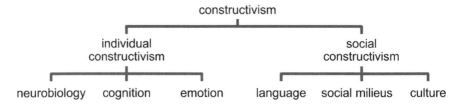

Fig. 10.2 Versions of constructivism

our everyday language[3] with its grammar and syntax is a collective memory, a reservoir of a variety of social historical experience. Language is a construction of reality and of social action. Language allows humans to find orientation and coordinate action (Fig. 10.2).

Gergen shows "how the meaning of our words does not depend on the characteristics of the world, but on their relationship to other words. Meaning first arises within texts or languages" (Gergen 2002: 59). With its lyric poetry the romantic movement was a major force in re-constructing the aesthetic quality of nature. A careful interaction with nature can be expressed in language but there can also be a linguistic environmental pollution. Language – as emphasized by social constructivism – does not reproduce reality, it interprets and creates realities (one thinks of Hölderlin, Trakl or Rilke). Language is closely related to communication and is an expression of traditions and feelings (Arnold and Holzapfel 2008).

The boundaries of our language point to the boundaries of our world. That also means that we must learn to speak about sustainability. The complexity of our language and the complexity of our (ecological) environment are mutually dependent on each other. Language contains visions of a socially and environmentally sustainable future. "As we describe and explain, so do we fashion our future" (Gergen 2002: 68). Gergen and Gergen (2003) thus pleads for a 'narrative pedagogy'. Biographical narratives are vivid and descriptive constructions and reconstructions of realities. Narratives are social construction processes that not only require a story-teller but also listeners who question, supplement and correct the story. Narratives are social confirmation in that they link the individual and the unique with the common, the consensual. Sustainability communication can be revived by a new learning culture of social-ecological story-telling.

Conclusion

Constructivism is not a theory that explains how the world is created. Constructivism is more a meta-theory that explains why the question as to the nature of the world cannot be satisfactorily answered. Constructivism is thus – following Niklas

---

[3] In German the expression *Umgangssprache* is revealing. We might loosely translate it as 'interaction language', the language that we use to 'deal' with each other and with the world.

Luhmann – a different type of theory. It makes a statement about the observer dependence of the theory itself and so an answer cannot be expected to the question of what sustainability communication is. However constructivism can show ways to think about sustainability communication – in the face of ambiguity and uncertainty, plurality and contingency.

References

Arnold, R., & Holzapfel, G. (Eds.). (2008). *Emotionen und Lernen. Die vergessenen Gefühle in der Erwachsenenpädagogik*. Baltmannsweiler: Schneider Hohengehren.
Balgo, R., & Werning, R. (Eds.). (2003). *Lernen und Lernprobleme im systemischen Diskurs*. Dortmund: Balser.
Ciompi, L. (2003). Affektlogik, affektive Kommunikation und Pädagogik. *REPORT Weiterbildung, 2003*(3), 62–70.
Gergen, K. (2002). *Konstruierte Wirklichkeiten*. Stuttgart: Kohlhammer.
Gergen, K., & Gergen, M. (Eds.). (2003). *Social construction: A reader*. London: Sage.
Lenzen, D. (1999). *Orientierung Erziehungswissenschaft*. Reinbek: Rowohlt-Taschenbuch-Verlag.
Luhmann, N. (1990). *Soziologische Aufklärung* (5th ed.). Opladen: Westdeutscher.
Roth, G. (2001). *Fühlen, Denken, Handeln*. Frankfurt am Main: Suhrkamp.
Simon, F. (Ed.). (1997). *Lebende Systeme*. Frankfurt am Main: Suhrkamp.

Part III
Practice of Sustainability Communication

Chapter 11
Climate Change as an Element of Sustainability Communication

Jens Newig

Abstract Climate change communication is a new and fast-developing element of sustainability communication. It can be conceived of both as communication about and as communication of climate change, referring to an analytical and a normative dimension of the concept. In this contribution, climate change communication will be outlined and its role within the larger field of sustainability communication will be discussed.

Keywords Climate change • Climate change communication • Mass media • Media discourse • Participation

Climate Change as a Sustainability Issue that Calls for Communication

Anthropogenic climate change constitutes a paradigmatic sustainability problem, reaching well beyond what is commonly known as 'environmental problems'. For three fundamental reasons, communication is a key element in societal strategies to cope with climate change.

First of all, the issue of climate change is characterised by a high level of complexity and uncertainty. As a global issue, its causes and (potential) implications vary greatly around the world (Young et al. 2006). The complex and highly non-linear system of interdependencies among the elements of the global climate are still not fully understood, rendering forecasts of the effects of greenhouse gas emissions and other factors highly uncertain, if not impossible. As moreover the

J. Newig (✉)
Research Group Governance, Participation and Sustainability,
Institute for Environmental and Sustainability Communication,
Leuphana University Lüneburg, Germany
e-mail: newig@uni.leuphana.de

stakes involved in climate change policy are so high, scholars such as Funtowicz and Ravetz (1993) call for new modes of science involving increased communication, dialogue and the involvement of stakeholders to broaden the information basis, but also to include broader societal values.

Second, while the urgency of climate change is generally acknowledged, there is not anything like a global consensus on what goal is to be achieved. Many countries have agreed on a goal to limit global warming to an increase of 2°C, but further consequences such as increased variability, extreme weather events or sea level rise affect different regions in such different ways that no agreement is in sight. Thus, the goals as to what extent climate change should be contained remain deeply ambiguous. Communication will play a key role in discussing possible and desirable emission targets as well as other normative aspects (Voß et al. 2007).

And third, the capacities to govern climate change and its causes are widely distributed among a great variety of societal actors on multiple levels of decision-making, making implementation of those few goals that have been agreed on all the more difficult. Once again, communication is advocated as a means of dealing with this dimension of the issue. In particular, network-like forms of co-ordination that enable effective argumentation, bargaining and social learning are regarded as conducive to governing climate change in the face of distributed action capacity (Voß et al. 2007; Newig et al. 2010).

All of these dimensions of climate change as a sustainability problem call for societal communication. When speaking of 'climate change communication' as one particular and fast-developing aspect of sustainability communication, two different perspectives can be taken – and are in fact taken. The first aspect regards communication *about* climate change. Important questions are: How and to what extent does society – and do societal subsystems and actors – communicate about the issue, what are its connotations, how is it framed, how is it linked to other issues? What options are discussed, e.g. mitigation or adaptation? Public discourse in the (international) public sphere and the role of the mass media are key aspects. This scholarly perspective is an analytical one. The second aspect regards the communication *of* climate change. Here, the perspective is more instrumental or managerial, and focused on a sender-receiver chain. Important questions are: How do those who know (or think they know) about climate change communicate it to others? This concerns, importantly, the role of science and of environmental groups who seek to educate others (e.g. politicians or the broad public) about climate change or the necessity to act in favour of its containment.

Both aspects of climate change communication are of course related and share common elements. From a normative perspective, sustainability communication can be viewed as a process of mutual understanding (Michelsen 2007).

Communication *About* Climate Change: The Societal Perspective

Climate change is a dynamic subject of communication. Different actors and social subsystems engage in climate-related debates on various levels from the very local to the global, and with different views and intentions. As an element of sustainability communication, climate change communication mostly takes place in public. This is not to say there is no private communication about it. Indeed, communication in networks among lay persons may form an important basis for societal communication about climate change.

In line with communication-based concepts in sociology (e.g. Luhmann 1995), climate change communication takes place in societal subsystems, the most important being the media, politics and science (Weingart et al. 2000). Discourse analysis is employed to characterise specific kinds of communication within these spheres, and how they 'irritate', or interact with, one another.

It was only in the course of the emancipation of the bourgeoisie from the aristocracy in England and France in the eighteenth century that a 'public sphere' emerged, involving the growing idea of free citizens with the right to form their own opinion of public affairs and to participate in the public process of opinion formation (Habermas 1981). In those days, the 'public' consisted largely of personal communication. Reaction times were quite long, and only certain groups could participate in it. It was not until the development of mass media – press, radio, television and later the internet – that a broader kind of public came into being by enabling ordinary citizens to receive (and disseminate) politically relevant information (McQuail 1994).

The mass media thus constitute an essential element of today's public sphere and serve an important function in public communication and discourse. Typically, media communication centres around particular issues and follows a logic inherent to the media system (Luhmann 1971). Regarding complex environmental and sustainability issues such as climate change, the mass media tend to ignore their inherent uncertainties, transforming them into a sequence of events leading to catastrophe and requiring immediate action (Weingart et al. 2000).

Clearly, climate change has become a highly important issue for the public sphere, also internationally. In quantitative terms, since autumn 2006 hundreds of articles have been printed per month in a single newspaper, the conservative German daily newspaper 'Frankfurter Allgemeine Zeitung' or F.A.Z. (Fig. 11.1). Although other Western countries report lower amounts of newspaper attention (e.g. Gavin 2009 for the UK), overall issue salience has shown an enormous rise. A distinctive theme 'career' can be identified (see Fig. 11.1).

In contrast to classic assumptions about media attention, the climate change issue does not show a distinct 'issue attention cycle' (Downs 1972). Rather, there was a build-up of attention that reached, for the time being, its peak in early 2007, months after two events of crucial importance: the October 2006 publication of the Stern Review Report (Stern 2007), holding that the economic costs of climate change exceed those of an effective mitigation many times over, and the publication of the spring 2007 report of the Intergovernmental Panel on Climate Change (IPCC 2007), likewise stressing the harmful effects of anthropogenic climate change. In addition, Al Gore's documentary 'An Inconvenient Truth' had a powerful impact when it was broadcast in May 2006 and (Egner 2007).

Media communications research provides a number of partly competing explanations for the immense rise in media attention. In a purely realist approach, the strength of public or media attention would mirror real-world events. In the case of

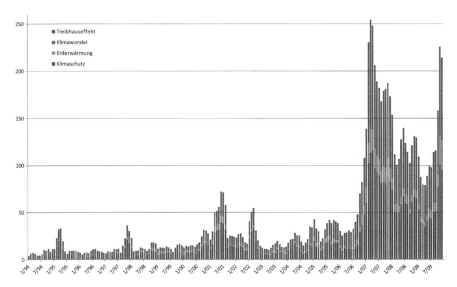

Fig. 11.1 Dynamics of news media communication about climate change. Depicted are numbers of articles per month in the German daily newspaper "Frankfurter Allgemeine Zeitung" for the period of 1994–2009, in which the keywords greenhouse effect (Treibhauseffekt), climate change (Klimawandel), global warming (Erderwärmung) or climate protection (Klimaschutz) appear (3 period moving average)

climate change, these would include an actual increase in average temperature, natural disasters attributed to it (such as Hurricane Katrina, which devastated New Orleans in August 2005) or other indicators. However, such events hardly predict media attention to climate change. Scholars following Downs (1972) see the ups and downs of media attention in the intrinsic quality of the issue itself, such as its more or less 'exciting' qualities or the facts of who would gain or lose when political action is taken. In a yet more constructivist perspective, the inherent logic of the media system and its narrative considerations are seen to play a crucial role as well (McComas and Shanahan 1999).

The media discourse on climate change has not only developed quantitatively, but also qualitatively. For instance, the discourse has moved from more scientific issues to those of mitigation and adaptation (see Weingart et al. 2000 for the early period from 1975 to 1995). Whereas climate change articles used to appear predominantly in the science section of newspapers such as the German F.A.Z., they have since moved to the politics section. Part of climate change communication in the media is the discussion of whether it is actually happening or not. In that it helped bring the issue to the political agenda and keep it there, media discourse has clearly had an important agenda-setting function (Pralle 2009).

The political system itself can, according to Luhmann, be further differentiated into the centre, formed by the government and the opposition, and the periphery, starting with political parties and interest groups and extending to social movements and the broad public ('Publikum').

Climate change communication in the political centre, as communicated by politicians, is essentially about the power of interpretation. In their analysis ranging from 1975 until 1995, Weingart et al. (2000: 270) illustrate how, "in political discourse, climate change was first constructed as humankind's all-embracing meta-problem and, in a later phase, was reconstructed and transformed into a problem of normal political regulation and routine". These dynamics were later partly reproduced, albeit with a greater level of urgency, in the political debates of the late 1990s and early 2000s, closely coupled to the media discourse outlined above. Similar to the media discourse, the political discourse initially framed climate change as a scientific issue but later it was differentiated and integrated into various sub-fields of public policy such as energy and transport (carbon emissions) or agriculture (nitrogen emissions). Yet in every phase, political communication never simply mirrored scientific discourse but took it up according to its own logic. For instance, one important function of the political system has been to reduce the complexity of scientific discourse by concentrating on CO_2 emission targets, thus stressing the relative autonomy of the dynamics of political climate change communication (Weingart et al. 2000).

Most prominently in the United States, political discourse has been divided into climate change 'believers' and 'sceptics', characterizing the issue as an intractable dispute (Fletcher 2009). Former US president George W. Bush declared: "No one can say with any certainty what constitutes a dangerous level of warming, and therefore what level must be avoided" (White House 2001). Years later, in the wake of the Stern Review Report (Stern 2007), a 'green transformation/opportunity frame' emerged in political climate change communication (Fletcher 2009).

Regarding the 'periphery' of the political system, the general public, two different modes of communication can be distinguished. First, large parts of public communication on climate change happen via the mass media. Indeed, media attention and media communication can be assumed to mirror public attention and communication (Newig 2004). Second, there is direct interpersonal communication. Clearly, knowledge about climate change issues has greatly increased in recent years (Nisbet and Myers 2007) and there is widespread concern about the phenomenon, but personal engagement still remains on a low level (Lorenzoni et al. 2007). Yet little is known about how people actually communicate with each other, through which means and how this communication is linked to the professional modes of the societal climate change communication discussed so far. Future research will have to employ network approaches in order to explore the role climate change plays in everyday communication.

Ultimately, the discourse on climate change originated in the sphere of science, with scientific discourse pointing to the policy relevance of its findings (Weingart et al. 2000). Within the scientific discourse, there has always been a strong emphasis on the uncertainties involved and the continuous need for further research efforts. Over time, perception of the global climate system has changed from a purely physical, chemical and biological one toward the notion of a system coupled with the human sphere in that it is affected by human action and in turn is a source of dangers for society (Weingart et al. 2000). Moreover, scientific discourse has introduced a distinction as to the two policy options in the face of climate change: adaptation and mitigation.

The IPCC defines mitigation of climate change as "an anthropogenic intervention to reduce the sources or enhance the sinks of greenhouse gases", whereas "adaptation to climate change refers to adjustment in natural or human systems in response to actual or expected climatic stimuli or their effects, which moderates harm or exploits beneficial opportunities" (IPCC 2001: 750).

To conclude, societal communication *about* climate change has reached a level of near omnipresence, taking place in different societal spheres in a variety of forms and highly dynamic with respect to changing connotations and framings.

Communication *Of* Climate Change: The Perspective of Governance and Education

Communication *about* climate change is of a discursive character. Society, or subsystems thereof, discuss a sustainability issue of high public relevance. Communication *of* climate change is different. Here, certain senders (seek to) convey their message to a certain receiver or audience. Climate change communication in this respect refers in particular to the efforts of science, environmental NGOs and other actors hoping to persuade policy-makers or the broader public of the urgency of climate change and the need to act accordingly. Thus understood, climate change communication becomes a part of risk communication, which can be defined as "communication intended to supply laypeople with the information they need to make informed, independent judgments about the risks to health, safety, and the environment" (Morgan et al. 2002: 4, see also Chap. 3). Specific purposes of climate change communication are to inform and educate individuals, to achieve some type and level of social engagement and action, and to bring about changes in social norms and cultural values (Moser 2010).

Since communication *of* climate change has clear intentions about its desired effects, it can – in contrast to communication *about* it – be assessed in terms of 'success'. Have the recipients been reached? Have they understood the message? Have they, perhaps, changed their values and changed their behaviour? The key question to be posed is how climate change can be communicated effectively in order to promote mitigation and/or adaptation.

Communication *of* climate change takes an elitist stance, making a central distinction between experts and laypersons in respect to their climate change related knowledge and capacities (Read et al. 1994; Bostrom et al. 1994; Nerlich et al. 2010). Scientists in particular "have long held and will continue to hold a privileged position as knowledge holders, messengers, and interpreters of climate change" (Moser 2010: 37). The perspective starts from a perceived need to educate the lay public (or professionals who are nevertheless 'lay' persons with respect to climate change). Several studies have revealed a severe lack of understanding even of basic principles of climate change and related causes on the part of the lay public (Bostrom et al. 1994; Sterman and Sweeney 2007).

The reasons for such a perceived lack of knowledge and understanding on the part of the broad public are manifold. It could be suspected that early 'climate change communicators' – mainly physical scientists and environmentalists without training in communication – did not communicate as effectively as they could have if they had had professional training (Moser 2010). More importantly, as has been stressed above, climate change is an enormously complex issue that is difficult to comprehend and to convey to others (Nerlich et al. 2010; Ockwell et al. 2009). The uncertainties involved in global circulation models are still immense, and while agreement even on the range of effects is difficult among scientists, it is even more challenging to communicate these to laypersons, let alone the implications of appropriate mitigation and adaptation responses. Unlike many other environmental problems, climate change can rarely be directly perceived through the senses, its causes are distant as are many of its impacts – both geographically and temporally. Therefore, "climate change – no matter how certain and urgent to experts – for now, and maybe for some time, is fundamentally a mediated, ambiguous problem for most audiences and easily trumped by more direct experiences" (Moser 2010: 36).

Given this sobering analysis, 'climate change communicators' and, increasingly, scholars of climate change communication have developed 'strategies' and 'tactics' on how to improve communication in order to better educate lay publics and initiate behaviour change. "Perhaps the most obvious communication strategy is the provision of information about climate change and the threat it poses, along with information about effective and practical responses. Another tactic is to stress the contribution of proposed climate policies to the achievement of other social and economic objectives such as energy security and employment. Messages aimed at citizens need to be simple and clear, which implies focusing on just a few selected indicators of climate change and its impacts, along with a small number of proposed solutions, and making use of metaphors and analogies to make it easier for citizens to understand complex ideas. Messages also need to be tailored to particular audiences and repeated as often as possible" (Compston 2009: 741).

In addition, Moser (2010) uses established insights of communication theory to make a strong case for a professionalisation of climate change communication. This includes a clear reflection on the motives and goals of communication, how messages are constructed and framed, and how they are conveyed. An important element is the type of language, the metaphors and images employed. For instance, environmental NGOs in the UK launched a number of climate action campaigns that tried "to prove that climate change is real through visual means", pointing to the dangers and vulnerabilities involved in the issue (Manzo 2010: 105). Well-known to a large audience is former US vice president Al Gore's documentary 'An Inconvenient Truth', conveying the narrative of immediate danger. Dominant images are endangered elements in non-human nature such as melting glaciers or endangered polar bears (Manzo 2010).

Notwithstanding its popularity, scholars of climate change communication also point to the limits of this 'public understanding of science' model, in which experts educate lay people (Nerlich et al. 2010). In particular, the dominant quest for behavioural change on the level of individuals, which has only had very limited success, is being increasingly questioned. Instead, the social and societal nature of behaviour comes to the fore. Recent communication efforts thus aim to generate social acceptance for 'low carbon' regulation or to stimulate grass-roots collective action, reconciling 'top-down' and 'bottom-up' approaches (Ockwell et al. 2009).

Climate Change, Sustainability Communication and Participation

Increasingly, the top-down, one-way mode of communication is questioned in favour of dialogue and discourse. Communication *of* climate change thus approaches the sphere of communication *about* it. Recently, serious failures in climate change communication have stunned public debate, such as the IPCC's erroneous scenario of Himalayan glaciers melting by 2035, which IPCC officials continued to uphold under doubtful circumstances. This contributed to declining public confidence in climate scientists (Leake and Hastings 2010).

Not only is the privileged position of science eroding; it is also increasingly acknowledged that the 'lay' public's perceptions differ fundamentally according to culture and context. "With this in mind, there is no such thing as an effective communication per se" (Nerlich et al. 2010: 106). Communication strategies, it is recommended, ought to take into account the different perceptions, views and interests of publics and policy-makers around the globe. Furthermore, communication of climate change and related discourses about it are clearly not neutral with respect to power issues. A multitude of actors with their own vested interests are trying to frame and shape climate change debates for their own benefit, making it a crucial issue of equity who communicates to which audience (Feindt and Oels 2005).

These different strands of argument ultimately call for a participatory approach to climate change communication as an important element of sustainability communication (Few et al. 2007). "It has to be acknowledged that civil society has an important function, alongside the market and state, as an instrument of steering in attaining sustainable development. Participation has to be understood as a another structural-policy instrument" (Michelsen 2007: 36). Given the dominant role of science in climate change discourse, participation can be important both to connect science with policy-making and, more traditionally, to policy-makers and stakeholders or the larger public (van den Hove 2000). A variety of methods are at hand (see Chap. 16), whose success in terms of sustainability gains, however, is not always uncontested (Newig and Fritsch 2009). Article 6 of the 1992 United Nations Framework Convention on Climate Change requires signatory states to promote and facilitate "public participation in addressing climate change and its effects and developing adequate responses". Scholars from social-ecological systems research call for participatory modes of communication so as to build adaptive capacity in order to cope with global sustainability issues such as climate change (Adger et al. 2005).

Climate change is one of the most pressing issues of sustainability, and climate change communication is an element of sustainability communication. Whereas 'sustainability' as a concept has predominantly been a topic of academic and elite discourse while remaining abstract for large parts of the public, climate change has successfully moved to the fore of public attention. Therein lies, if it is linked adroitly to other global issues of sustainability and sustainable development, the huge transformative potential of the issue. A particular opportunity lies moreover in the emerging global discourse on climate change. Should it prove possible to construct a global public sphere around the climate change issue, then the participation of civil society in this and other issues of sustainable development would be greatly facilitated.

References

Adger, W. N., Arnell, N. W., & Tompkins, E. L. (2005). Successful adaptation to climate change across scales. *Global Environmental Change, 15*, 77–86.

Bostrom, A., Morgan, M. G., Fischhoff, B., & Read, D. (1994). What do people know about global climate change? 1. Mental models. *Risk Analysis, 14*(6), 959–70.

Compston, H. (2009). Networks, resources, political strategy and climate policy. *Environmental Politics, 18*(5), 727–46.

Downs, A. (1972). Up and down with ecology – The "issue-attention cycle". *The Public Interest, 28*, 38–50.

Egner, H. (2007). Surprising coincidence or successful scientific communication: How did climate change enter into the current public debate? *Gaia-Ecological Perspectives for Science and Society, 16*(4), 250–54.

Feindt, P. H., & Oels, A. (2005). Does discourse matter? Discourse analysis in environmental policy making. *Journal of Environmental Policy & Planning, 7*(3), 161–73.

Few, R., Brown, K., & Tompkins, E. L. (2007). Public participation and climate change adaptation: Avoiding the illusion of inclusion. *Climate Policy, 7*(1), 46–59.

Fletcher, A. L. (2009). Clearing the air: The contribution of frame analysis to understanding climate policy in the United States. *Environmental Politics, 18*(5), 800–16.

Funtowicz, S. O., & Ravetz, J. R. (1993). Science for the post-normal age. *Futures, 25*(7), 739–55.

Gavin, N. T. (2009). Addressing climate change: A media perspective. *Environmental Politics, 18*(5), 765–80.

Habermas, J. (1981). *The theory of communicative action: Reason and the rationalization of society* (Vol. 1). Boston: Beacon.

IPCC. (2001). Annex B: Glossary of terms. In J. J. McCarty, O. F. Canzianni, N. A. Leary, D. J. Dokken, & K. S. White (Eds.), *Climate change 2001: Impacts, adaptation, and vulnerability – contribution of Working Group II to the third assessment report of the Intergovernmental Panel on Climate Change*. Cambridge: Cambridge University Press.

IPCC. (2007). *Climate change 2007. The physical science basis* (Contribution of working group I to the fourth assessment report of the Intergovernmental Panel on Climate Change). Cambridge: Cambridge University Press.

Leake, J. & Hastings, C. (2010). *World misled over Himalayan glacier meltdown*. The Times Online, 17 January 2010. Retrieved July 30, 2010, from http://www.timesonline.co.uk/tol/news/environment/article6991177.ece.

Lorenzoni, I., Nicholson-Cole, S., & Whitmarsh, L. (2007). Barriers perceived to engaging with climate change among the UK public and their policy implications. *Global Environmental Change-Human and Policy Dimensions, 17*(3–4), 445–59.

Luhmann, N. (1971). Öffentliche Meinung. In N. Luhmann (Ed.), *Politische Planung* (pp. 9–34). Opladen: Westdeutscher.

Luhmann, N. (1995) *Social systems*. Palo Alto: Stanford University Press (Original edition, Soziale Systeme. Grundriß einer allgemeinen Theorie, 1984).

Manzo, K. (2010). Imaging vulnerability: The iconography of climate change. *Area, 42*(1), 96–107.

McComas, K., & Shanahan, J. (1999). Telling stories about global climate change: Measuring the impact of narratives in issue cycles. *Communication Research, 26*(1), 30–57.

McQuail, D. (1994). *Mass communication theory: An introduction* (Vol. 3). London: Sage.

Michelsen, G. (2007). Nachhaltigkeitskommunikation: Verständnis – Entwicklung – Perspektiven. In G. Michelsen & J. Godemann (Eds.), *Handbuch Nachhaltigkeitskommunikation: Grundlagen und Praxis* (pp. 25–41). München: Ökom.

Morgan, M. G., Fischhoff, B., Bostrom, A., & Atman, C. J. (2002). *Risk communication: A mental models approach*. Cambridge: Cambridge University Press.

Moser, S. C. (2010). Communicating climate change: History, challenges, process and future directions. *WIREs Climate Change, 1*(1), 31–53.

Nerlich, B., Koteyko, N., & Brown, B. (2010). Theory and language of climate change communication. *WIREs Climate Change, 1*(1), 97–110.

Newig, J. (2004). Public attention, political action: The example of environmental regulation. *Rationality and Society, 16*(2), 149–90.

Newig, J., & Fritsch, O. (2009). Environmental governance: Participatory, multi-level – and effective? *Environmental Policy and Governance, 19*(3), 197–214.

Newig, J., Günther, D. & Pahl-Wostl, C. (2010). Synapses in the network. Learning in governance networks in the context of environmental management. *Ecology & Society, 15*(4).

Nisbet, M. C., & Myers, T. (2007). Twenty years of public opinion about global warming. *Public Opinion Quarterly, 71*(3), 444–70.

Ockwell, D., Whitmarsh, L., & O'Neill, S. (2009). Reorienting climate change communication for effective mitigation forcing people to be green or fostering grass-roots engagement? *Science Communication, 30*(3), 305–27.

Pralle, S. B. (2009). Agenda-setting and climate change. *Environmental Politics, 18*(5), 781–99.

Read, D., Bostrom, A., Morgan, M. G., Fischhoff, B., & Smuts, T. (1994). What do people know about global climate change? 2. Survey studies of educated laypeople. *Risk Analysis, 14*(6), 971–82.

Sterman, J. D., & Sweeney, L. B. (2007). Understanding public complacency about climate change: Adults' mental models of climate change violate conservation of matter. *Climate Change, 80*, 213–38.

Stern, N. (2007). *The economics of climate change: The Stern review*. Cambridge: Cambridge University Press.

van den Hove, S. (2000). Participatory approaches to environmental policy-making: The European Commission Climate Policy Process as a case study. *Ecological Economics, 33*, 457–72.

Voß, J.-P., Newig, J., Kastens, B., Monstadt, J., & Nölting, B. (2007). Steering for sustainable development: A typology of problems and strategies with respect to ambivalence, uncertainty and distributed power. *Journal of Environmental Policy & Planning, 9*(3–4), 193–212.

Weingart, P., Engels, A., & Pansegrau, P. (2000). Risks of communication: Discourses on climate change in science, politics, and the mass media. *Public Understanding of Science, 9*(3), 261–83.

White House. (2001). *President Bush discusses global climate change,* 11 June [online]. Retrieved July 30, 2010, from http://georgewbush-whitehouse.archives.gov/news/releases/2001/06/20010611-2.html.

Young, O. R., Berkhout, F., Gallopin, G. C., Janssen, M. A., Ostrom, E., & van der Leeuw, S. (2006). The globalization of socio-ecological systems: An agenda for scientific research. *Global Environmental Change, 16*, 304–16.

Chapter 12
Biodiversity and Sustainability Communication

Maik Adomßent and Ute Stoltenberg

Abstract Biodiversity can be seen as an exemplary issue for sustainability communication. In addition the conflictual relationship between conservation and sustainable use will be illustrated using selected examples. From the perspective of successful sustainability communication, this chapter will show not only the complexity of cause and effect but also the options there are to conserve biological diversity. Special importance is attributed to the systematic relationship between biological and cultural diversity, since this is given a key role in the formulation of recommendations for developing sustainability communication.

Keywords Biodiversity • Biological diversity • Cultural diversity • Human-nature relationships • Sustainability communication

Background

Biodiversity as a decisive factor in economic, social and cultural development and biodiversity as the integrity of an intact natural world make this topic a central issue in sustainable development. It is a problematic field with such a variety of causal interrelationships that it can be seen as exemplary for networked thinking, a skill that is crucial for shaping the future responsibly. 'Conservation and sustainable use' – two principles that are contested in current political strategy debates – are connected with economic interests, cultural values and global distributive

M. Adomßent (✉)
Institute for Environmental and Sustainability Communication,
Leuphana University Lueneburg, Scharnhorststrasse 1, D-21335 Lueneburg, Germany
e-mail: adomssent@uni.leuphana.de

U. Stoltenberg
Institute for Integrative Studies, Leuphana University Lueneburg,
Scharnhorststrasse 1, D-21335 Lueneburg, Germany

justice and are thus an example for the negotiation of sustainability principles. Biodiversity can thus be seen as an exemplary area for the problems facing sustainability communication.

With the ratification of the Convention on Biological Diversity in 1992, 191 signatory countries have so far underlined the importance of this issue, making it one of the most important conservation and sustainability agreements in the world. In 2002 the partners to this convention pledged to make a notable reduction in the loss of biodiversity by 2010. This goal has not been achieved; the ninth Conference of Parties (COP) in 2008 was used as an occasion for a number of countries to step up their activities. Currently 107 countries have developed National Biodiversity Strategies and Action Plans (NBSAPs), a further 23 parties to the agreement were asked to initiate corresponding measures by 2010. Germany has fulfilled its obligations arising out of signing the CBD, which it ratified in 1993, and produced a 'National Strategy for Biological Diversity' (BMU 2007). In order to increase public awareness of the topic of biological diversity and its many aspects of communication and education, the United Nations has declared 2010 to be the International Year of Biodiversity.

The diversity of life and the spatially specific qualities of nature are not new objects of fascination. In illuminated medieval manuscripts realistic illustrations of field flowers show the close attention paid to the domestic 'little nature'. Profusely illustrated volumes of baroque garden flowers show the diversity of flowers found in these gardens and go beyond a purely biological interest in the taxonomy of plants, although these botanic gardens did in fact have their origin as collections of biological diversity representing the systematisation of the plant kingdom and making a contribution to knowledge about the species. Human intervention in nature through breeding was not undertaken alone through considerations of utility, but was motivated – as can be seen in the variety of forms and colours of tulips or roses – by aesthetic (and arguably also economic) reasons. And finally conservation and the founding of conservation organisations have their roots in an engagement for particular natural areas or species.

Sustainable development is a global vision that has led to a change in thinking about biodiversity. It can no longer be seen primarily from an ecological or aesthetic perspective but it is now a factor for sustainable development in a number of central fields of action. And these are decisive for the quality of the future. Climate change, as caused by the industrial production and processing of food, the type of land use, the use of pesticides and synthetic fertilisers together with habits of consumption, is closely related to imminent losses of biodiversity. This has made the use of biodiversity for global food production, medical and technological knowledge, for the development opportunities of countries of the southern hemisphere a crucial issue (Fig. 12.1).

Biodiversity is considered – similar to sustainable development – as too vague a term for communication processes and as a result 'biological diversity' is used in its place (Kitchin 2004). The definition of the Convention on Biological Diversity shows its advantage in clarifying the primary importance given to the intimate relationship between species diversity, genetic diversity and the conservation of

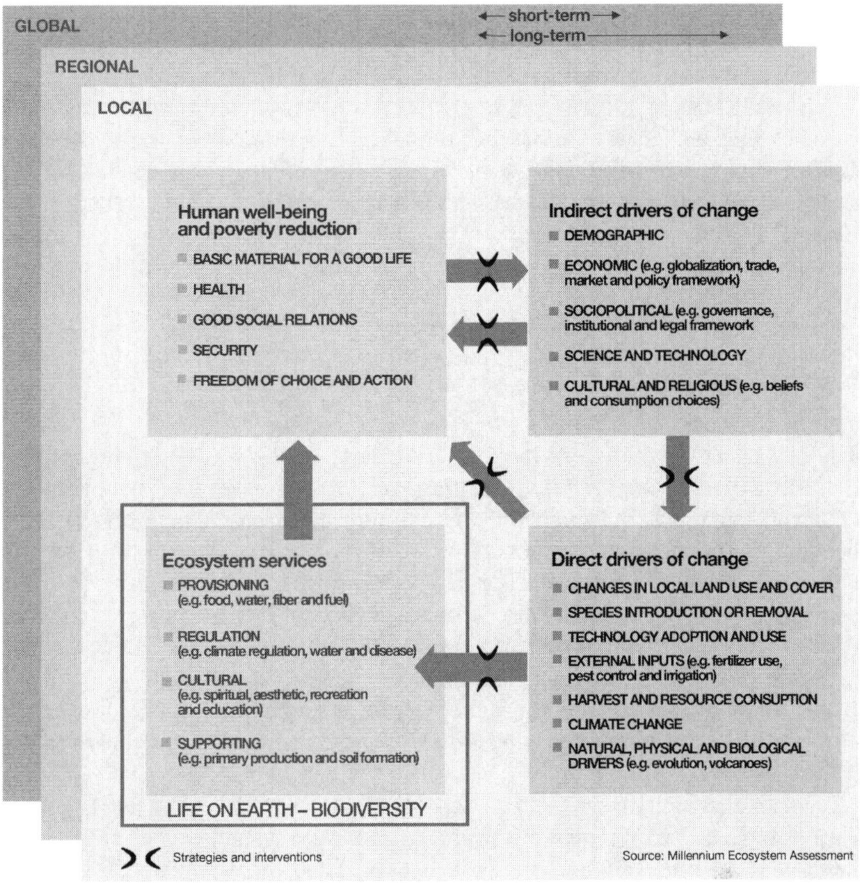

Fig. 12.1 Interactions between biodiversity, ecosystem services, human well-being, and drivers of change (Source: MA 2005: iii)

ecosystems, not the conservation of individual species alone. This leads to one of the most important messages, namely that also from a purely natural science perspective what is important are the systematic relationships. The following section uses a number of examples to illustrate such human-nature interrelationships and the tension between the conflicting priorities of conservation and use.

Examples of Causal Relationships

Unsustainable practices in ways of living and economic practices have led to a global loss of species that has reached a level as much as 1,000 times the natural rate (MA 2005: 3f.). In the twentieth century 30% of all vertebrates have become extinct.

A number of ecosystems – including the oceans, which had once been almost impossible to imagine as being affected by human activity – have been fundamentally disrupted and even destroyed.

These phenomena can be ignored or considered part of an unpredictable natural world so long as humans are not directly affected by the consequences. There are many natural interrelationships associated with a loss of biodiversity that are able to arouse interest, even among people with different social and cultural backgrounds. However it is more likely that biodiversity will receive more attention when individuals can connect such interrelationships with a desirable life or with specific interests.

Biodiversity and Food

The concept of agrobiodiversity provides general access to the problem of biodiversity, because food security concerns everyone, whatever their age, social or cultural background. The number of different cultivated plants in the world can only be estimated. There are tens of thousands of different types of wheat, corn, rice and potato. However estimates show that generic diversity is now 75% less than at the beginning of the twentieth century. This means that an ever increasing number of people is dependent on an ever decreasing number of species and breeds, which moreover originate from more or less the same genetic material. Five types of grain (wheat, corn, rice, barley and millet (also known as sorghum)) account for over half of total human consumption, and 95% of all plant-based foodstuffs come from just 30 species (FAO 2005).

Global interrelationships, such as securing world food supplies through the use of adapted regional varieties, may not be appreciated by everyone. A better way of communicating the value of biodiversity is to show its effect on daily food consumption. The loss of diversity in species and in plant types not only affects the flavour of food but also its healthfulness (when important plant compounds are lost).

Biodiversity and Seeds

For thousands of years, genetic diversity has been a guarantee that – under a variety of environmental conditions and without the use of external means of production – crops could be harvested in a sustainable fashion, offering protection against the widespread outbreak of diseases and providing a degree of food security. In countries in the southern hemisphere this is still the core of a stabile and sustainability-oriented agricultural and land use system. Diversity provides the security necessary for survival by partially compensating for a loss of crops due to adverse conditions (e.g. drought). By contrast in industrial countries the focus is on breeding genetic characteristics that promise high yield crops. In order to breed qualities that are as uniform as possible (e.g. synchronous harvest times), sexual reproduction is prevented

Table 12.1 Demand for water in plant and animal production (Source: Pimentel et al. 1997)

Water use (in l) for production of 1 kg of	
Potatoes	500
Wheat	900
Alfalfa	900
Sorghum	1,100
Corn	1,400
Rice	1,910
Soya beans	2,000
Poultry	3,500
Beef	100,000

by the use of hybrid varieties, even though this is known to increase their vulnerability, for example to pathogens. As a result classic plant cultivation involves breeding disease-resistant varieties. This resistance is however often quickly broken down. A race against time evolves that leads to a lack of genetic variability both within a given variety of plant (homogeneity) and between different varieties (relatedness) (FAO 1996).

Biodiversity and Consumption

The relationship between biodiversity and consumption does not need to be reduced to food – although this would involve the greatest opportunities to move consumers towards a more sustainable lifestyle and preserve biodiversity. In industrial countries production, processing and marketing often use as much as ten times the energy as the product itself contains (EEA 2009: 34ff.).

Against a background of striving to achieve greater distributive justice, it is evident that the world population cannot be fed using the current standards of food production in industrial countries. Especially the production of meat wastes precious resources, as can be seen in the demand for both energy and water (Table 12.1).

Biodiversity and Climate Change

Such unsustainable production and consumption patterns contribute to climate change, which is one of the most important factors leading to the loss of biological diversity (MA 2005: 9). Neither of these global phenomena can be analysed separately. The effects of climate change expected to occur in Europe will most probably take the form of losses in biodiversity. A decrease in the area of agricultural land and Mediterranean wooded areas is to be feared as is a dramatic reduction in wetlands, which play a critical role as CO_2 sinks (EEA 2010a). Surprisingly, negative impacts from climate-related increases in temperature on species populations are forecast not only for temperate zones but also for tropical regions (Wright et al. 2009).

Biodiversity and Tourism

The tourism and leisure industry is one of the fastest growing economic sectors worldwide. For many emerging economies it offers an important source of hard currency and jobs, as well as less dependence on other economic sectors. Natural habitats with higher levels of biological diversity are increasingly important to tourist activities and nature-related offerings have become a significant growth segment of the tourist industry. Paradoxically through fast and more or less uncontrolled growth, tourism can also have the effect of destroying the environment and so contribute to the loss of local identities and traditional cultures (Wilde and Slob 2007).

However tourism, especially nature-related travel, has considerable potential for contributing to the conservation and sustainable use of biological diversity. Income can be used for the conservation of natural resources, with sustainable tourism making a contribution to economic development particularly of remote regions (Vancura 2008).

Biodiversity and Land Use

As one of the greatest threats to biodiversity is the use of land for housing development and transport infrastructure, it is essential to make the conservation of biodiversity an integrated task of urban development and comprehensive spatial planning.

Sustainability communication can make use of research findings on new methods of construction that take account of social, economic and cultural aspects. Other concepts involve securing the survival of flora and fauna through the use of biocorridors, for example across highways, through cooperation in the spatial planning of biotope networks and through the alternative use of green spaces. For urban areas green axes and watercourses can be planned to run through built up areas. But also the quality of urban green spaces must be reconceived, by cultivating neighbourhood gardens with agricultural plants or replacing biodiversity-poor park lawn areas with domestic trees and shrubs (Müller et al. 2010).

Biodiversity and Wilderness

With the exception of five high-biodiversity wilderness areas world-wide, high levels of biological diversity are not necessarily found in a given wilderness area, and so the goals of biodiversity and those of wilderness conservation are not congruent. However even if, following Mittermeier et al. (2003), barely 20% of plants and 10% of terrestrial vertebrate animals are endemic in wilderness areas (such as Amazonia, the Congo, New Guinea, the Miomba-Mopana woodlands and the North American deserts), these refuges play an important role in a global perspective, including as a control variable for measuring the health of our planet.

Furthermore, especially the African wilderness areas are crucial refuges for cultural diversity, in which a large number of indigenous languages and religions are preserved (Pretty et al. 2009).

Biological and Cultural Diversity and Its Communication

Biological diversity in cultural landscapes, especially agrobiodiversity, is a result of cultural processes. Humans have bred and colonised the plants and animals that were best fitted to the living conditions in a particular environment. With their meadows, hedgerows and field borders, cultural landscapes are rich in diverse varieties and species of flora and fauna. In fact even in the rainforest, there are more medical plants where humans have selectively logged individual trees and built trails than in primary forest. The "culturalisation of nature" (Küster 1995: 370) and the diversity of human ways of life make a direct contribution to biodiversity.

Cultural identity and biological diversity are closely related (Pretty et al. 2009). Foods made from regional agricultural products or wild plants and animals and served in season or on particular occasions give individuals a feeling of belonging to a region or to a group. Slow Food, an organisation that is regionally anchored and at the same time internationally active, uses this knowledge for its engagement in preserving biodiversity. Cultural customs and rituals often make use of flora and fauna from the surrounding area and so serve to confirm group identity. Excellent examples here are trees, which are part of rituals in many parts of the world. Cultural practices are a guarantee for their conservation and so also for their environment.

The ruthless degradation of biodiversity is a result of European expansion into the southern hemisphere, colonization and the exploitation of natural resources, but also more recently by technological developments, for example the excessive use of nitrogen and phosphorous nutrients or the promotion of monocultures and the concentration on a small number of animal species by the seed and food industry (Scherr and McNeely 2008). This also has consequences for cultural diversity, as it indirectly impinges on the basis for its existence.

In turn cultural homogenisation and the disappearance of traditional ways of life accelerate the loss of biodiversity. There is a loss of knowledge for example of how to cultivate plants in a particular micro-climate (e.g. the Alps) or of old varieties of vegetables or of the use of wild plants (FAO 2005). Accelerated by new cultural practices brought about by mass tourism and mass production, this development has over a number of decades led to a radical reduction and a comparatively small number of domestic species and varieties of vegetable foods (FAO 1996; Thrupp 2000). The same holds for domestic livestock. When time and personal relationships and the quality of animal foods no longer play a role in the relationship between humans and animals, then certain species will no longer be kept (TGRDEU 2010).

Cultural diversity is thus not only to be seen from a perspective of cultural products and forms of expression that are a common heritage of humankind to be preserved (UNESCO 2008). It is also a condition for the conservation of biodiversity – and not only in relationship to indigenous people living in rainforests, whose natural world together with themselves is threatened. In order to become aware of and attend to these relationships, a number of different instruments and initiatives have been developed on both regional and international levels. NGOs and government programmes have developed concepts to support indigenous peoples (see for example Mars and Hirschfeld 2008). On a regional and community level international gardens and neighbourhood gardens are practical initiatives and at the same time opportunities for communication about sustainable development. Community or government programmes or grassroots initiatives for the conservation of old cultivated plants are the global answer to the weakening of food security and the quality of life.

Biodiversity as an Element of Sustainability Communication

Sustainability communication cannot limit itself to informing or educating the populace about complex ecological relationships. It would be an important step if information about biodiversity were not provided in a purely textual form, but instead would be related to everyday contexts or to a variety of areas of social experience. Such strategies must be supplemented by developing possibilities to preserve biological diversity. The complex relationships surrounding biodiversity, as shown above, offer a good opportunity. There are many potential actors. The questions for sustainability communication include:

- Who are the major actors?
- What opportunities are there for them?
- What types of cooperation are possible in a common field of action?

Science has an important role to play here. For example, DIVERSITAS, a global association of actors in biodiversity research, has the goal of supporting the search for ways to a sustainable use of world-wide biotic resources. This could involve findings in conservation psychology (Corbett 2006; Manfredo 2008) as well as further social science research in the advising of political decision-makers in matters concerning biodiversity (Gilbert et al. 2006). Finally inter- and transdisciplinary research projects can show opportunities to take action that have a real chance of being put into practice (www.biostrat.org).

Biodiversity is a problem area that was initially seen by the public to be largely global in context, i.e. biodiversity as an issue connected with the rainforests. There is a factual reason for this as rainforests have the greatest density of biodiversity and probably also the greatest treasure of species and genetic diversity. But for Europeans the rainforest is also a fascinating, exotic, mystical region, which is not necessarily considered to be in the realm of actual possibilities to take action (Flitner 2000; Gallup Organization 2007). NGOs that are engaged in protecting the rainforest and showing specific actions that can be taken there have an important role to play in sustainability communication (e.g. www.oroverde.de).

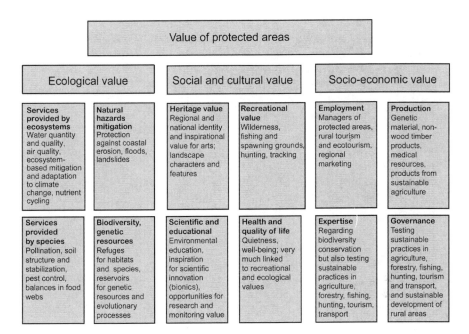

Fig. 12.2 Ecological, social and socio-economic values of protected areas (Source: EEA 2010b)

Opportunities to take action can also be found in classic nature conservation, which can also be involved in sustainability communication (Rientjes 2000). National parks and other protected areas can be used as examples of biodiversity and create a relationship for individuals to this issue. Environmental associations that involve their members and others in monitoring actions (for example bird censuses in a number of countries) provide opportunities for public engagement. From a sustainability point of view biosphere reservations are very good subjects for sustainability communication, as locations for finding ways of life that harmonise biodiversity and business (see Fig. 12.2).

The conservation of biodiversity must not however be limited to protected areas. Cultivated landscapes are a challenge for the conservation and possibly also the development of biodiversity. Sustainability communication can make use of these relationships, showing how both biotope and species and genetic diversity are a necessary element of culture (UNESCO 2008). The example of the water cycle in the high plains of Ecuador and Peru shows how sustainability communication can accompany sustainability development (Rivadeneira et al. 2009). The human relationship to water is a cultural product. Colonial influences have led to a 'forgetfulness of water'. A more sensitive relationship to water, the careful development of agro-cultures is experienced as the stabilisation of cultural and biological system. It creates an awareness of ecosystem services, food security and biodiversity.

Alliances at a regional level need to be found that are capable of organising sustainability communication as a process of communication. This includes farmers wanting to use older varieties of seed and resisting the planting of genetically modified seeds (FAO 1996). The linking of biodiversity with taste, cultural heritage, aesthetics and the efforts to preserve the creators of biodiversity, even on a small scale, is a concept of sustainability that can unite consumers, producers, the catering industry and educational institutions (Pokorny 2009). An example of such an alliance is Terra Madre, a global network of farmers, cooks and universities and research institutes (www.terramadre.org).

A more fundamental argument involves understanding biodiversity as a 'source of knowledge and information' to be used creatively and productively.[1] Bionics is a new branch of knowledge and industry together with bio-architecture can make a contribution to sustainable development and can give new cultural impulses as well as awaken more interest in the conservation of biodiversity. However there is a danger that companies will make use of this knowledge from nature without pursuing a complex sustainability strategy and cultural diversity. Sustainability communication is then challenged to expose such economic and political structures and contribute to an understanding of how they affect ecosystems and the quality of human life.

References

BMU (Federal Ministry for the Environment, Nature Conservation and Nuclear Safety) (Ed.). (2007). *National strategy on biological diversity*. Berlin. Retrieved March 26, 2010, from http://www.bmu.de/files/pdfs/allgemein/application/x-download/national_strategy_biodiv.pdf.
Corbett, J. B. (2006). *Communicating nature: How we create and understand environmental messages*. Washington, DC: Island Press.
EEA. (2010a). *10 messages for 2010. Climate change and biodiversity*. Copenhagen: EEA.
EEA. (2010b). *10 messages for 2010. Protected areas*. Copenhagen: EEA.
EEA/European Environment Agency. (2009). *Progress towards the European 2010 biodiversity target*. Copenhagen: EEA.
FAO. (2005). *Building on gender, agrobiodiversity and local knowledge*. A Training Manual. Rome, Italy.
FAO/Food and Agriculture Organization of the United Nations. (1996). *Global plan of action for the conservation and sustainable utilisation of plant genetic resources for food and agriculture*. Leipzig, Germany.
Flitner, M. (Ed.). (2000). *Der deutsche Tropenwald. Bilder, Mythen, Politik*. Frankfurt am Main: Campus.
Gallup Organization. (2007). *Flash Eurobarometer Series #219. Attitudes of Europeans towards the issue of biodiversity*. Survey conducted by The Gallup Organization Hungary upon the request of Directorate-General Environment.
Gilbert, K., Hulst, N., & Rientjes, S. (2006). *SoBio: Social science and biodiversity. Why is it important? A guide for policymakers*. Tilburg: ECNC – European Centre for Nature Conservation.
Kitchin, T. (2004). *'Assuring Biodiversity': A brand-building approach*. The Glasshouse Partnership. Retrieved March 25, 2010, from http://www.glasshousepartnership.com/downloads/branding.pdf.

---

[1] This is a formulation from the new Ecuadorian constitution of 2008; See Plan Nacional para el Buen Vivir 2007–2010, p. 132.

Küster, H. (1995). *Geschichte der Landschaft in Mitteleuropa von der Eiszeit bis zur Gegenwart.* München: C. H. Beck.

MA/Millenium Ecosystem Assessment. (2005). *Ecosystems and human well-being. Biodiversity synthesis.* Washington, DC: World Resources Institute.

Manfredo, M. J. (2008). *Who cares about wildlife? Social science concepts for exploring human-wildlife relationships and conservation issues.* New York: Springer.

Mars, E. M., & Hirschfeld, M. (Eds.). (2008). *Der Wald in uns.* München: Oekom.

Mittermeier, R. A., Mittermeier, C. G., Brooks, T. M., Pilgrim, J. D., Konstant, W. R., da Fonseca, G. A. B., & Kormos, C. (2003). Wilderness and biodiversity conservation. *Proceedings of the National Academy of Sciences of the United States of America, 100*(18), 10309–10313.

Müller, N., Werner, P., & Kelcey, J. G. (2010). *Urban biodiversity and design* (Conservation science and practice). Weinheim: Wiley-VCH.

Pimentel, D., Houser, J., Preiss, E., White, O., Fang, M., Mesnick, L., Barsky, T., Tariche, S., Schreck, J., & Alpert, S. (1997). Water resources: Agriculture, the environment, and society: An assessment of the status of water resources. *BioScience, 47*(2), 97–106.

Plan Nacional para el Buen Vivir (PNBV) 2007–2010 and 2009–2013. Retrieved May 11, 2011, from http://www.senplades.gov.ec.

Pokorny, D. (2009). Marketing agrobiodiversity: Rhön Biosphere Reserve, Germany. In S. Stolton (Ed.) Communicating values and benefits of protected areas in Europe. *BfN-Skripten, 260*, 53–58.

Pretty, J., Adams, B., Berkes, F., Ferreira de Athayde, S., Dudley, N., Hunn, E., Maffi, L., Milton, K., Rapport, D., Robbins, P., Sterling, E., Stolton, S., Tsing, A., Vintinnerk, E., & Pilgrim, S. (2009). The intersections of biological diversity and cultural diversity: Towards integration. *Conservation and Society, 7*(2), 100–112.

Rientjes, S. (Ed.). (2000). *Communication nature conservation. A manual on using communication in support of nature conservation policy and action.* Tilburg: ECNC – European Centre for Nature Conservation.

Rivadeneira, S., Suárez, E., Téran, J. F., & Velásquez, C. (2009). *Gente y ambiente de páramo: realidades y perspevtivas en el Ecuador.* Quito: Editiones Abya-Yala.

Scherr, S. J., & McNeely, J. A. (2008). Biodiversity conservation and agricultural sustainability: Towards a new paradigm of 'ecoagriculture' landscapes. *Philosophical Transactions of The Royal Society of Biological Sciences, 363*, 477–494.

TGRDEU/Central Documentation on Animal Genetic Resources in Germany. (2010). Retrieved March 1, 2010, from http://tgrdeu.genres.de/default/gefaehrdung/index/?lang=en.

Thrupp, L. A. (2000). Linking agricultural biodiversity and food security: The valuable role of sustainable agriculture. *International Affairs, 76*(2, Special Biodiversity Issue), 265–281.

UNESCO. (2008). *Links between biological and cultural diversity-concepts, methods and experiences. Report of an International Workshop.* Paris: UNESCO. unesdoc.unesco.org/images/0015/001592/159255e.pdf.

Vancura, V. (2008). Tourism service providers as partners for the conservation of biodiversity – the PAN parks sustainable tourism strategy. In BMU (German Federal Ministry for the Environment, Nature Conservation and Nuclear Safety), BfN (German Federal Agency for Nature Conservation) & INOEK (Institute of Outdoor Sports and Environmental Science (Eds.), *Biodiversity and sport – prospects of sustainable development.* Series "Outdoor Sports and Environmental Science", *23*, 69–74.

Wilde, J., & Slob, B. (2007). Eco-holidays – The sustainable tourism paradox. *ClimateChangeCorp climate news for business.* Retrieved March 26, 2010, from http://www.climatechangecorp.com/content.asp?ContentID=4780.

Wright, J. S., Muller-Landau, H. C., & Schipper, J. (2009). The future of tropical species on a warmer planet. *Conservation Biology, 23*(6), 1418–1426.

Chapter 13
Communicating Sustainable Consumption

Lucia A. Reisch and Sabine Bietz

Abstract This chapter provides a general overview of sustainable consumption as an object of politics and research. Specifically it looks into the challenges to develop communication campaigns that will motivate a broad public to engage in more sustainable lifestyles. An emotional, experience-oriented, mass-medial communication concept based on an "ecotainment" strategy can successfully increase interest and lower acceptance barriers towards sustainable consumption in the general population. As a case of good practice, findings of a communication project that aimed at increasing the public's attention to sustainable consumption issues are presented.

Keywords Consumption • Communication campaigns • Sustainable lifestyle • 'Project Balance' • Ecotainment

Sustainable Consumption as an Object of Politics and Research

Sustainable consumption has been coined 'the underdog' in the arena of sustainability initiatives (Kolandai-Matchett 2009). While conceptual approaches and empirical evidence on sustainable consumption issues have rapidly increased over the years (see Clark 2006), efforts to communicate this concept have received only little political attention in most countries worldwide.

L.A. Reisch (✉)
Department of Intercultural Communication and Management,
Copenhagen Business School, Copenhagen, Germany
e-mail: lr.ikl@cbs.dk

S. Bietz
Institute for Consumer Behaviour & European Consumer Policy, SRH Hochschule Calw,
University of Applied Sciences, Lederstr. 1, 75365 Calw, Germany

In Germany, for instance, the importance of private consumption for the development of a more 'sustainable Germany' has been a topic of discussion inside expert circles since about the mid-1990s. Consumer and environmental advocacy groups, through conferences and action campaigns, were the first to attempt to popularise the idea of sustainable consumption. While concepts such as curbing climate change, rescuing the rain forests and saving energy have meanwhile been mainstreamed in most European societies – and are even used as 'good causes' in cause-related marketing by businesses – the consumer movement is still the key driver, both in its breadth and depth, for sustainable or 'strategic' consumption. In addition to traditional modes of communication, activists and advocacy groups are making increasing use of interactive communication paths and new media such as social networks and other Web 2.0 applications to reach their publics (Hinton 2009; Repo et al. 2009). As opposed to traditional one-way 'expert to consumer' communication, the latter allow for instant reaction, multi-sender communication, and consumer-produced up-to-date content and participatory approaches, empowering consumers and promoting consumer interest with a formerly unknown momentum (Reisch 2010).

The perception of the importance of consumers as powerful agents of change, in both politics and research, has fundamentally changed as engaged consumers are seen as a source of creativity, competence, seriousness and potential pressure. The share of organic, fair trade and sustainable products in the market is now growing. Only a decade ago, consumption was not seen as a relevant topic for either politics or economic research. Traditionally, economic, environmental and even development policy research focused on the company, on production processes and on the value chain. The extension of interest to the demand side of the market and 'sustainable production and consumption patterns' was long overdue and a welcome new perspective (Reisch et al. 2010).

Consumption can be defined as a complex multi-level process of acquisition, use and disposal of co-production and self-production in households and social networks. It has now become a relevant dimension of sustainability politics. Today the promotion of sustainable consumption and production patterns is internationally established as a research field and is starting to be institutionalized in the form of university chairs, research institutes, university courses and research programmes. In the political sphere – stimulated by international conferences from Rio to Marrakech to Copenhagen – sustainable consumption has gradually been accepted as an area of political activity on all levels, from regional, national, European to international and supranational. In summer 2008, the European Commission (2008) published an 'Action Plan on Sustainable Consumption and Production' (SCP). Its main target is to arrange a dynamic framework to improve the energy and environmental performance of products and encourage their uptake by consumers. The Action Plan consists of three parts: smarter consumption and environmentally better products, leaner production and global markets for sustainable products. The character of the Action Plan is that of a 'communication' of proposed measures and activities, which are then to be implemented by specific actions, as stipulated by directives and regulations. Member states have started – and partly implemented – their own SCP activities in different ways. Some national governments have elaborated their own concepts in action plans and policy programmes (Czech Republic, Finland, Hungary, Poland and the United Kingdom), whereas in other states (Austria, France, Italy, Malta, the Netherlands) SCP is embedded in larger national strategies for sustainable development. Some member states pursue approaches that focus on single policy instruments, such as greener public purchasing or eco-labelling (e.g. Denmark, Germany) without an explicit policy framework document (IÖW et al. 2009).

While there is a steadily growing base of research on sustainable consumption as well as increasing political consensus, there is still an urgent need to learn more about how to successfully communicate messages to consumers as a general population and to hard-to-reach specific socio-economic and lifestyle groups (the young, the old, hedonists, the less educated etc.). To date, there is not a conclusive theory of sustainability communication. Instead different approaches have been developed that are not integrated into a systematic effort. The broadest approach used is the social marketing concept (Frame and Newton 2007; Golding 2009; Peattie and Peattie 2009; McKenzie-Mohr 2000), which has proved successful in the public health area.

Sustainable Consumption in Different Areas of Basic Needs

The context of relevant everyday structures together with an account of the social milieu, lifestyle, life situation, attitudes and values yield potential starting points for communication campaigns oriented to changing consumer behaviour (OECD 2008). At the same time it shows an overall view of a particular goods supply system, and also shows the innovation potential on the supply side as well as the potential of civic institutions such as consumer organizations or the media to disseminate information and educate the general public.

The European Environmental Agency (EEA 2005: 14) observes that while there has been some progress "(…) the general trend is an increase in environmental pressures, because consumption growth is outweighing gains made through improvements in technology". According to this report, private households are directly responsible for one-fourth of final energy use and two-thirds of municipal waste generation in the EU. A study on behalf of the Directorate-General for the Environment identified three main areas contributing to about 70–80% of environmental pressures, namely food/drink, housing and private mobility (Tukker et al. 2006). On the other hand, through their purchasing decisions, households can influence the market penetration of 'socially sound' products. For instance, in the United Kingdom, fair trade labelled products have a 5% market share of tea, a 5.5% share of bananas, and a 20% share of ground coffee (Golding 2009).

By using intelligent technologies and new materials in production processes, from design-for-environment strategies in product design and alternative service and use concepts, there are great opportunities to achieve ecological savings and gains simply through efficiency. These however will only be completely realised if the technological advance is complemented by a change – cultural and social – in use behaviour. A large number of important durable consumer goods such as cars or washing machines contribute as much as 80% to the total environmental impact caused by use. According to a British study (Ventour 2008), one-third of the food bought by UK households ends up as waste, 61% of which could have been eaten if it had been better managed. However, communicating the optimum 'use regime' to the various target groups is not an easy undertaking, as can be seen in the 'eco top ten innovations' (Grießhammer et al. 2007).

In the basic needs area of clothing the culturally conditioned behaviour of 'fast-changing fashion' results in an amount of 11 kg of little-used clothing per person, per year. It would hardly be a promising undertaking to launch a communication campaign against a post-modern consumption society defined by fast paced changes in fashion. Representatives of a consistency strategy are therefore proponents of the 'pleasurable' consumption of 'intelligent' materials and attractive designs, which are materially unproblematic 'from cradle to cradle', in production, use and disposal, e.g. compostable (Braungart and McDonough 2002).

The communication of sustainable consumption is especially problematic in the basic needs areas of living and mobility not only because individual mobility and homeownership are 'leading cultural goods', greatly exaggerated symbols that help to determine an individual's social standing, but also because the architectural infrastructure (building code, transport infrastructure plans etc.) all too often turns out to be a barrier that can only be overcome at considerable, and prohibitive, personal expense. Finally, social and ecological goals are often contradictory. Consider for example the 'democratisation' of long-distance travel and the discussion about discount airlines. Public statements by experts that are not clearly structured, or even contradictory, lead to a loss of credibility that makes the communication of more sustainable consumption alternatives even more problematic.

In the basic needs area of food there has been considerable success with the strategy of buying regional, seasonal and organic products. Common-pool benefits (quality of the environment, regional development, regional value creation, etc.) and individual benefits (health, taste etc.) often coincide and alliances of motive can thus be plausibly communicated. The organic food branch is booming like no other, one of the drivers in Germany being its sale by large-scale discounters. If this market is to develop it is crucially important that communication campaigns reach infrequent 'test' buyers and induce them to habitualise their choice.

Things become more difficult when new technologies for sustainable consumption show an ambivalent effect. In the area of green genetic technology, there are also experts who argue that genetically modified seeds and foods have the potential to reduce world hunger. Another contested field is nanotechnologies. Similar ambivalences can be found in the area of communication technology. While the internet drastically sinks the costs of information, organisation and contracts (i.e., so-called 'transaction costs'), increases transparency and makes it possible for the poorer 'world consumers' to collectively articulate their opinion, at the same time the production, use and networking of information technology itself is extremely material and energy intensive (Reisch 2003).

There is not enough empirical data to demonstrate how behaviour in a particular basic needs area relates to other areas, e.g., whether there are so-called 'wedge behaviours' that allow for a 'foot-in-the-door' strategy to other consumption fields. Also, it is still empirically unclear how powerful 'spillover effects' between different areas of consumption are (Thøgersen and Crompton 2009), let alone how such effects could be systematically exploited in communication strategies. Moreover, in spite of years of research, there is little valid knowledge about the so-called consumer 'attitude-behaviour gap' (Young et al. 2009). That attitudes have only a very moderate effect on actual behaviour has been well studied in social psychology and is now a well-known fact about human behaviour. The dominant explanation for the divergence between attitude and behaviour is the 'low-cost hypothesis', i.e., consumers live up to their attitudes when the perceived costs of such a choice are low. Yet, the costs argument has been supplemented by the 'high-justice hypothesis', which takes into account intrinsic motivation and moral imperatives such as the principle of fairness (Coad et al. 2009; Montada and Kals 1995). Behavioural economics and economic psychology have empirically shown that the quest for 'fair deals' is at the heart of most human transactions (Thaler and Sunstein 2008). This would mean that once the consumer has understood the necessity of sustainably oriented action he or she would not need any further private benefits or material incentives to choose more sustainable service packages. There is still however a need for fair conditions. In many cases the consumer makes a conditional commitment, based on certain ideas about justice, but will only act upon it when certain actions have been taken in advance by other consumers, by the government or by companies. Once these conditions have been fulfilled, they still have to be effectively communicated.

A further reason why environmental and social awareness has little relevance for consumer behaviour is that consumers feel that the government and business bear a greater responsibility for environmental and development problems than they do. This is compounded by consumers being subject to the 'illusion of marginality'. This type of illusion is familiar to environmental and risk psychology, but it is also highly relevant for the conceptualisation of successful sustainable consumption communication strategies. Furthermore, recent research in behavioural economics and sustainable choice has shown that consumer behaviour is much more dependent on the stimuli and barriers of the immediate choice contexts (e.g., the stimuli and information provided at the point of sale; the accessibility, affordability and availability of sustainable alternatives in a neighbourhood) and is influenced to a far greater extent by human biases and heuristics than has been assumed in consumer science. Experimental and survey evidence implies that a smart-choice architecture making use of 'choice editing' (Yates 2008) and 'sustainable defaults' (e.g. green power as the 'default' electricity supply or healthy food served in school canteens unless another choice is exercised) is at least as effective in 'nudging' consumers into more sustainable choices as are efforts to influence consumers' knowledge, preferences and attitudes via communication tools (Thaler and Sunstein 2008).

Potentials and Pitfalls, Options and Barriers of Sustainable Consumption

Due to the generally high level of material prosperity and the possibilities to access a great variety of goods and services on offer in Western consumer societies, there is a critical mass of consumers who have a considerable degree of discretion in their purchasing decisions. This still holds true in times of economic crisis. In spite of a certain amount of path dependence in consumption, resulting from structural 'lock-in effects' as well as budget and availability limitations, many consumers would be able to choose more or less environmentally and socially friendly alternatives in the individual phases of the consumption process – from reflection to determining needs, from deciding whether to buy, rent or exchange, from use and maintenance to disposal and recycling. However on a behavioural level they meet with barriers or restrictions that systematically impede sustainable consumption behaviour. These are often prohibitively high additional costs – which are often also considered to be unfair – contradictory information signals, opportunistic supplier behaviour and structural overloading (Yates 2008). Only a small and especially committed group of consumers will do the 'right' thing in such 'wrong' structures. If communication is to be successful it must take account of these options and restriction. The abundance of empirical research on factors influencing behaviour 'options' can be interpreted against this background.

A number of studies has shown that along with a positive attitude and concern for the environment and just conditions of action, the following factors have a supportive effect on sustainable consumption behaviour: being aware of having a variety of possibilities to act (option attractiveness and avoidance reaction); unambiguous knowledge relevant to the consequences of action (information about the costs of prosperity); economic incentives and disincentives (to the extent that these do not undermine intrinsically motivated behaviour); positive consumption experiences with ecological and socially fair services (regarding functional aspects of quality, but also concerning aesthetics, haptics and appearance); social recognition and moral regard arising from a consumption decision and as given by relevant reference groups and the social network; normative appeals where descriptive norms (what people typically do) are communicated together with injunctive norms (what people typically approve or disapprove of) (Cialdini et al. 2006); target group specific tailoring and framing of messages (Pelletier and Sharp 2008); visualising positive consequences to behaviour by technological or communicative feedback mechanisms (reward effect) such as 'smart meters'; and finally tailored and target group specific communication (environmental education, advice and information) especially using unconventional, emotionalising communication strategies as well as the targeted integration of informal – real and virtual – social networks and communities as intensifying or defensive 'communication buffers'.

Communication Promoting Sustainable Consumption Patterns

The increased efforts to popularise sustainable consumption do not appear to have had the success once hoped for. A qualitative jump beyond 'more of the same' might result from a reflection on the theoretical premises of sustainability communication. In fact many of the findings of media and communication sciences have been taken up by communicators of sustainability and many of their efforts can be considered state of the art. These include the following elements: messages should appeal to benefits and motivation that are either potentially personal or specific to a particular type of consumption (health, fitness, taste experiences, savings, convenience, wealth in time, social recognition etc.) instead of simply doing without consumer goods; it should be viable in everyday life, i.e. understandable and easy to put into practice as well as containing any needed service elements. The timing of the message also plays a role. In certain personal life transitions, such as the birth of a child or following an illness, the willingness to change habitual behaviour seems to be higher. From a technical perspective messages should be target-group specific and use a variety of media, including more visual media such as film, television and the internet and the formats successfully used in such media. The unspecified use of the prefixes eco and environmental now has a negative connotation and should be avoided, with positively connoted 'brand names' being used instead. While fear-provoking images certainly have a place in sustainability communication, they must be used selectively, and with caution. If people do not feel ecological and social threats are significant issues, using fear-provoking images is likely to cause denial, apathy and avoidance (O'Neill and Nicholson-Cole 2009).

According to the "ecotainment" concept (Lichtl 1999), mass medial entertainment concepts can be a suitable approach, especially when addressing consumer types who are otherwise critical of sustainability or disinterested. Emotion, experience and entertainment precede knowledge and will power. Aesthetics and surprise effects are part of an approach that uses art and culture as a medium of communication for sustainable development. The idea is that artists will offer their creative, aesthetic and artistic skills for the development of a culturally based sustainability discourse and spread the topic of sustainability through sensual-aesthetic experiences.

Especially these newer approaches to medialising sustainability have potential, because they are better suited to a post-modern, visually-oriented image and consumption culture of the 'generation Y' than traditional 'cognition-biased' formats (Jansson 2002). Multi-sensual formats are far more effective in allocating a particular symbolic meaning to products and consumption practices, which can then be decoded and evaluated by the target groups. This is a basic function of communication from a symbolic interactionism perspective. In a time of information overload and consumer confusion, these formats and media have a greater chance of being perceived in the first place. Without doubt there will still be groups that show no interest at all in the topic of 'sustainable consumption' or that have taken an explicitly defensive posture.

This can be related to a generally negative attitude toward consumption work, such as status-related consumption habitus, upwardly mobile consumption aspirations or everyday overload. The limits to communication become apparent at this point.

Meanwhile, a small number of mediated communication campaigns promoting mainstream 'sustainable consumption' have been described in the consumer research literature, employing theories on effective and persuasive communication (Kolandai-Matchett 2009). One of the early endeavours of this kind was a 2004 research project called 'Project Balance', which tested the potential of a mass media sustainable consumption communication format (Reisch and Bietz 2007; Reisch et al. 2010). 'Project Balance' was designed as a transdisciplinary joint research project, received funds from the German Ministry of Research and involved partners in consumption behaviour research, marketing research and media sciences working together with experienced practitioners from the field of media. Its central research question was whether and how an emotional, experience-oriented, mass-medial communication concept based an 'ecotainment' strategy could contribute to the dismantling of interest and acceptance barriers to sustainability topics, especially sustainable consumption among the general public (Schwender et al. 2008). The project developed a target group, situation-specific and TV-based communication concept to significantly increase the attention and interest of the general public in the concept of sustainability. Following the guidelines of sustainability communication outlined above, 'Project Balance' sought to be positive and entertaining and to motivate mainly by allying motives and personal benefits. The vocabulary 'eco' and 'sustainability' was never used explicitly, although of course the substance of the message was about sustainability. The emphasis was placed on technological innovation, health, wellness and nature. In a cross-media approach, the TV clips were supported by an internet service and podcasts.

The effects of the broadcast on consumer attitudes, knowledge and behaviour were studied by accompanying consumer research. In summary, results showed that the sustainability topic was very well received in this medial 'package' and that it generated positive attitudes and behavioural intentions in the targeted group of consumers who were 'little interested' in sustainability. Hence, the concept of highly emotional and subtle sustainability education clips wrapped in a popular 'science-light TV programme' might be a good choice to popularise sustainable consumer behaviours. For the future, it would be promising to develop 'cross-media' concepts that involve cooperation with Web 2.0 communities and eventually mobile internet applications to increase the up-to-datedness and usefulness of information at the point of purchase.

Sustainable consumption remains a central challenge for state and society. Policy-makers must initiate, stimulate and monitor the development of consumption so that both the natural limits of the planet and social equality are respected. In addition to other policy instruments, mass-mediated consumer communication is a key tool to inform, advise, stimulate and motivate consumers. Along with state-sponsored programmes, societal 'sub-political' actors, including the media and consumers themselves, can not only participate in reaching this goal, but can actually set the agenda, select the means and produce the contents themselves – in Web 2.0 applications.

References

Braungart, M., & McDonough, W. (2002). *Cradle to cradle: Remaking the way we make things.* New York: North Point Press.

Cialdini, R. B., Demaine, L., Sagarin, B. J., Barrett, D. W., Rhoads, K., & Winter, P. L. (2006). Managing social norms for persuasive impact. *Social Influence, 1*(1), 3–15.

Clark, G. (2006). Evolution of the global sustainable consumption and production policy and the United Nations Environmental Programme's (UNEP) supporting activities. *Journal of Cleaner Production, 15*, 492–498.

Coad, A., de Haan, P., & Woersdorfer, J. S. (2009). Consumer support for environmental policies: An application to purchases of green cars. *Ecological Economics, 68*, 2078–2086.

European Commission. (2008). *Communication from the commission to the European Parliament, the Council, the European Economic and Social Committee and the Committee of the Regions on the sustainable consumption and production and sustainable industrial policy action plan* (SEC (2008) 2111). Brussels: EU.

European Environment Agency (EEA). (2005). *Household consumption and the environment* (EEA Report No 11/2005). Copenhagen: EEA.

Frame, B., & Newton, B. (2007). Promoting sustainability through social marketing: Examples from New Zealand. *International Journal of Consumer Studies, 31*, 571–581.

Golding, K. M. (2009). Fair trade's dual aspect: The communications challenge of fair trade marketing. *Journal of Macromarketing, 29*(2), 160–171.

Grießhammer, R., Graulich, K., Bunke, D., Eberle, U., Gensch, C. -O., Möller, M., Quack, D., Rüdenauer, I., Zangl, S., Götz, K. & Birzle-Harder, B. (2007). *EcoTopTen – Innovationen für einen nachhaltigen Konsum (Hauptphase).* Öko-Institut e.V. in Kooperation mit Institut für sozial-ökologische Forschung (ISOE). Frankfurt/Freiburg: IOSE & Öko-Institut e.V.

Hinton, E. D. (2009). *Changing the world one lazy-assed mouse click at a time.* Working Paper 16: Environment, Politics and Development Working Paper Series. London: King's College, Department of Geography.

Institute for Ecological Economy Research [IÖW], Institute for European Studies – Free University of Brussels [IES-VUB] & National Institute for Consumer Research [SIFO]. (2009). *Innovative approaches in European sustainable consumption policies* (ASCEE). Final Report. Retrieved July 30, 2010, from http://www.ioew.de/home/downloaddateien/IOEW-SR_192_Approaches_Sustainable_Consumption.pdf.

Jansson, A. (2002). The mediatization of consumption: Towards an analytical framework of image culture. *Journal of Consumer Culture, 2*(1), 5–31.

Kolandai-Matchett, K. (2009). Mediated communication of 'sustainable consumption' in the alternative media: A case study exploring a message framing strategy. *International Journal of Consumer Studies, 33*, 113–125.

Lichtl, M. (1999). *Ecotainment: Der neue Weg im Umweltmarketing.* Vienna: Ueberreuter.

McKenzie-Mohr, D. (2000). Promoting sustainable behaviour: An introduction into community-based social marketing. *Journal of Social Issues, 56*(3), 543–554.

Montada, L., & Kals, E. (1995). Perceived justice of ecological policy and pro-environmental commitments. *Social Justice Research, 8*(4), 305–327.

O'Neill, S., & Nicholson-Cole, S. (2009). "Fear won't do it": Promoting positive engagement with climate change through visual and iconic representations. *Science Communication, 30*(3), 355–379.

OECD. (2008). *Promoting sustainable consumption: Good practices in OECD countries.* Paris: OECD.

Peattie, K., & Peattie, S. (2009). Social marketing: A pathway to consumption reduction? *Journal of Business Research, 62*, 260–268.

Pelletier, L. G., & Sharp, E. (2008). Persuasive communication and proenvironmental behaviours: How message tailoring and message framing can improve the integration of behaviours through self-determined motivation. *Canadian Psychology, 49*, 210–217.

Reisch, L. A. (2003). Potentials, pitfalls and policy implication of electronic consumption. *Information and Communications Technology Law, 12*(2), 93–109.
Reisch, L. A. (2010). Von blickdicht bis transparent: Konsum 2.0. In S. A. Jansen & N. Stehr (Eds.), *Transparenz*. Wiesbaden: VS Verlag für Sozialwissenschaften.
Reisch, L. A., & Bietz, S. (2007). How to convince the inconvincible? A mass mediated approach to communicate sustainable lifestyles to a low-interest target group. *International Journal of Innovation and Sustainable Development, 2*(2), 192–200.
Reisch, L. A., Spash, C. L., & Bietz, S. (2010). The socio-psychology of achieving sustainable consumption: An example using mass communication. In C. L. Spash, R. Holt, & S. Pressman (Eds.), *Post-Keynesian and ecological economics* (pp. 178–199). Northampton, MA: Edward Elgar.
Repo, P., Timonen, P., & Zilliacus, K. (2009). Alternative regulatory cases challenging consumer policy. *Journal of Consumer Policy, 32*(3), 289–301.
Schwender, C., Mocigemba, D., Otto, S., Reisch, L. A. & Bietz, S. (2008). Learning from commercials – communicating sustainability issues to new audiences. Why emotions matter. In T. Geer Ken, A. Tukker, C. Vezzoli, & F. Ceschin (Eds.), *Proceedings of the SCORE! Conference "Sustainable Consumption and Production: Framework for Action: Refereed Sessions III–IV"* (pp. 353–372). Paper presented at the Conference of the Sustainable Consumption Research Exchange (SCORE!) Network, Brussels.
Thaler, R. H., & Sunstein, C. R. (2008). *Nudge: Improving decisions about health, wealth, and happiness*. New Haven: Yale University Press.
Thøgersen, J., & Crompton, T. (2009). Simple and painless? The limitations of spillover in environmental campaigning. *Journal of Consumer Policy, 32*(2), 141–163.
Tukker, A. Huppes, G., Guinée, J., Heijungs, R., de Koning, A., van Oers, L., Suh, S., Geerken, T., Van Holderbeke, M., Jansen, B., & Nielsen, P. (2006). *Environmental impact of products (EIPRO)*. Analysis of the life cycle environmental impacts related to the final consumption of the EU-25. Brussels: Technical Report EUR 22284 EN.
Ventour, L. (2008). *The food we waste*. Banbury/Oxon: WRAP.
Yates, L. (2008). Sustainable consumption: The consumer perspective. *Consumer Policy Review, 18*(4), 96–99.
Young, W., Hwang, K., McDonald, S. & Oates, C. J. (2009). Sustainable consumption: Green consumer behaviour when purchasing products. *Sustainable Development* published online, Wiley InterScience (www.interscience.wiley.com). DOI: 10.1002/sd.394.

Chapter 14
Corporate Sustainability Reporting

Christian Herzig and Stefan Schaltegger

Abstract This chapter introduces the goals and benefits that are motivating companies to report on their sustainability activities and provides an overview of the historical development of sustainability reporting over recent decades. The presentation of challenges in sustainability reporting is followed by a critical appraisal of approaches to overcome these problems. The authors suggest a double-path approach which combines the strategic inside-out approach of performance measurement and management with the outside-in approach of adopting to the external requirements and conclude with consequences for the field of sustainability communication.

Keywords Benefits and goals of sustainability reporting • Historical development of sustainability reporting • Challenges in and approaches to sustainability reporting • Outside-in and inside-out perspective

Introduction

Recent decades have witnessed an exponential growth in non- or extra-financial reporting such as environmental, social or sustainability reporting. Mainly large companies, but also SMEs, are informing their stakeholders more and more often about their social and environmental performance through print-based reporting or their websites. Sustainability reporting, as the most comprehensive and integrative form of corporate reporting, has also gained attention among industry bodies and

---

C. Herzig (✉)
International Centre for Corporate Social Responsibility, Nottingham University Business School, Nottingham, UK
e-mail: christian.herzig@nottingham.ac.uk

S. Schaltegger
Center for Sustainability Management, Leuphana University Lueneburg, Lueneburg, Germany

associations, government institutions, consulting firms, non-governmental organisations and research institutions. At both national and international levels this can be seen in the increasing number of general and sector-specific frameworks and guidance documents, regulatory disclosure and reporting requirements and the interest of a variety of institutions in analysing and observing developments in sustainability reporting.

Goals and Benefits of Sustainability Reporting

There are many typologies of rationales that have been created to explain the existence of sustainability reporting. Explanations as to what motivates sustainability reporting include variants of accountability, legitimacy, stakeholder and political economic theories (Deegan 2002; Gray et al. 1995; Roberts 1992; Ullmann 1985; Tinker et al. 1991). As Buhr (2007) notes, these rationales can be closely interlinked and employed together as a way for a company to understand its reporting situation. Spence and Gray (2007) explored the motivations underlying social and environmental reporting in the UK. Perceived benefits and pressures, as observed by Spence and Gray, range from business efficiency, market drivers, reputation and risk management, stakeholder management, internal champions and mimetic motivations – each can be seen as expressions of ideas in the legitimate mores of the business and forming part of some overall business case.

Defining strategies to disclose sustainability information can be a way to gain, maintain and repair *legitimacy* (Deegan 2002). This applies to the public acceptance of the company generally, as well as to the acceptance of particular management decisions and activities by the company's key stakeholders.

Another explanatory motive underlying sustainability reporting can be the enhancement of a company's *reputation and risk management* (Bebbington et al. 2008). Outstanding corporate reputation is often related to higher brand value and may contribute to increasing business success (e.g. Fombrun 1996). In particular, reputation may be enhanced by reporting about successful engagement in non-market matters, i.e. in social and environmental projects that are not considered to be part of core business activities.

Reporting non-financial corporate activities signals a willingness to communicate about and deal with societal issues, and may serve to secure a continuing good relationship with the company's stakeholders. Companies that are perceived as being simultaneously high performers both in the market and for society may face less friction and problems in their business relationships with suppliers, traders, public authorities and other stakeholders. As a result companies can try to gain a *competitive advantage* in comparison to other companies that do not engage in sustainability activities or that do not communicate their achievements effectively enough. Besides *external benchmarking with competitors* or reporting leaders, companies may use *company-internal benchmarking* processes and systems to compare business units, production sites, etc. In this context, sustainability reporting can play a key role in creating transparency about responsibilities and accountability for activities and performance benchmarking.

Finally, with the collection and analysis of information as well as the creation of greater transparency, sustainability reporting can support internal information and control processes. Seen as a learning rather than an adaptive process, sustainability reporting may also initiate processes to enhance employees and manager awareness and motivation, and lead to individual and organisational changes that foster organisational performance. This requires critically reflexive processes where accepted rules, strategies and norms are questioned and improved (Gond and Herrbach 2006).

Which of these goals and benefits motivate management most to deal with sustainability reporting depends on the company-specific situation, on industry and market conditions, as well as on stakeholder constellations and management preferences. Moreover, since reporting initiatives convey a picture of corporate responsiveness to key societal concerns, they have changed over the last decades. Their historical development is described next.

Development of Sustainability Reporting

When reviewing the historical development of sustainability reporting over recent decades, it becomes apparent that corporate non- or extra-financial reports have changed perspectives and directions in response to different societal challenges (see Herzig and Schaltegger 2006). Figure 14.1 illustrates the different stages and forms of reporting with a particular focus on Europe using the three-pillar approach to sustainable development.

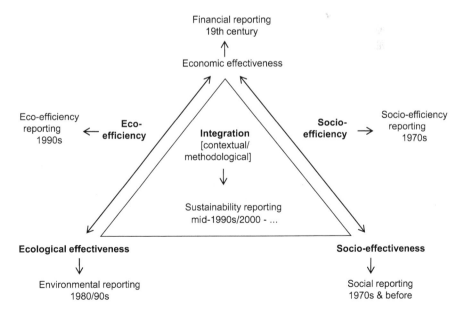

Fig. 14.1 Perspectives of sustainable development and development of sustainability reporting

Financial reporting originated in the nineteenth century and focuses on monetary principles and measures ('economic effectiveness' in Fig. 14.1). In the 1970s as income levels rose, the focus of society and politics shifted to objectives like quality of life, while at the same time the negative effects of quantitative economic growth and a Tayloristic organisation of production processes were seen critically in most parts of Europe. This led to a number of companies starting to publish their social goals, activities and impacts in specific *social reports*. This type of corporate reporting is often considered to be the first stage of non- or extra-financial reporting, although it has been preceded by the disclosure of employee and community issues within annual reports for many decades (Guthrie and Parker 1989). The essential concern of the reporting of *social balances* and the publication of social reports was to inform stakeholders about the company's activities, products and services, and related positive and negative social impacts *(socio-effectiveness)*. However by the end of the 1970s, social reports had become rare. Among the reasons for the decline were an inadequate target group orientation; the mismatch between the information interests of most stakeholders and social reports that were often scientifically designed and remote from the reality of most people's lives; the instrumentalisation of social reporting as a public relations tool, which reduced its credibility; the insufficient integration of social and financial reporting; and the positive economic and political development of Europe, with job movements to the services sector and improved working conditions (Dierkes and Antal 1985; Hemmer 1996).

About a decade later, in the late 1980s and early 1990s, *environmental reporting* emerged and to a large extent superseded early social reporting activities. One of the main aims of environmental reporting is to provide information on *ecological effectiveness* or, in other words, the absolute level of corporate environmental impacts such as air and water emissions, types and amounts of wastes, etc. *Environmental reporting* can be seen as a response to *hazardous incidents and environmental disasters* such as Schweizerhalle (Switzerland), Icmea Ltd. (Italy) and Hoechst AG (Germany) in the 1990s. In consequence, companies were perceived to be the major creators and causes of environmental problems. They started – partly forced by new laws (compulsory reporting), partly voluntarily – to provide information about environmentally relevant corporate activities to a variety of stakeholders. Until the end of the millennium, the number of environmental reports and the attention they received in the media and society increased significantly, and their average quality improved – from being primarily green glossaries and one-off reports in the beginning to more comprehensive environmental reports published on a regular basis. An example for a voluntary approach to environmental reporting is the *European Union Eco-Management and Audit Scheme (EMAS)*. It recognises companies that manage and improve their environmental performance and document their respective achievements using public *environmental statements*, a specific form of an environmental report.

In addition and sometimes exceeding these rather one-dimensionally oriented communication activities, reporting started to focus on *two-dimensional links between the economic and the environmental dimensions (eco-efficiency)* or – more rarely – the *link between the economic and the social dimensions (socio-efficiency)*. Since the mid-1990s, companies have increasingly disclosed information about the interrelationship between economic output and ecological input (eco-efficiency) in their environmental, business and financial reports. The concept of eco-efficiency, first developed in academia (Schaltegger and Sturm 1990), has been popularised by the World Business Council for Sustainable Development (Schmidheiny 1992), which subsequently took the lead in disseminating the eco-efficiency approach into business practice. In contrast to the history of the eco-efficiency concept, an analogous analysis and presentation of socio-efficiency, as the link between social and economic issues, has received less attention in business reports – partly due to difficulties in quantifying social aspects. Socio-economic considerations were however already present in the 1970s and social reporting practices and elements of these, such as the value added statements, have survived in sustainability reports (e.g. Diageo's 2009 Corporate Citizenship Report, The Co-operative's Sustainability Report 2009) or financial reports (e.g. the 2009 Annual Reports of BMW and Merck).

Since the mid-1990s, and increasingly towards the end of that decade, attention shifted to sustainability reports (e.g. Kolk 2004). These reports reflect companies' claims to depict an overall picture of their sustainability activities and to inform stakeholders as to what extent and how corporations contribute to sustainable development. One of the main challenges related to such integrative sustainability reporting is to outline the impacts of corporate activities from the different angles of the three (environmental, social and economic) perspectives, including conflicting goals, dilemmas, synergies, priorities and decision-making processes (*contextual integration challenge*). In practice these aspects have so far been considered primarily in an *additive and less than integrative manner* – failing to recognise and mention possible and actual conflicts and challenges embodied in their approach to corporate sustainability (Gray 2006; Herzig and Godemann 2010). In addition, sustainability reporting requires reflection on how to incorporate principles and aspects derived from the vision of sustainability development, in particular those of social justice, intra- and intergenerational equity, pluralistic and consultative decision-making, and different temporal timeframes – something that Buhr (2007) criticises as largely underdeveloped in current practice. Nevertheless, unlike in the 1970s, social aspects within sustainability reports are nowadays often *more globally* and also *more comprehensively* dealt with, in terms of moral and ethical questions of sustainable development, such as child labour in the supply chain, human rights, poverty alleviation, gender issues, trading relationships, etc. Besides the contextual challenge, integrative sustainability reports also face a *methodological integration challenge* as the different forms of existing reports, further communication activities and channels, and the underlying information management and accounting approaches that provide the reporting information need to be interwoven.

Companies are currently attempting to integrate environmental, social and financial accounting information in very different ways. Three main sustainability reporting strategies can be distinguished:

- *Distinctive stakeholder- and theme-specific reports*: One reporting strategy is the publication of a series of different company reports such as environmental reports, environmental statements, social reports or corporate citizenship reports.

Each of these deals with specific aspects of corporate sustainability and addresses different stakeholder groups.
- *Stand-alone sustainability reports*: In this reporting strategy, companies publish stand-alone sustainability reports that provide information about the company's ecological, social and economic sustainability activities and performance, often following the format of earlier environmental reports and published in addition to financial reports. One of the earliest examples is the so-called 'Triple P-Report' (People, Planet and Profits) of Shell, published in 1999, whose title already indicates its three-dimensional reporting character.
- *Extended financial reports and integrated (business) reports*: Selected environmental (and social) aspects of corporate performance have received more attention in financial reports in recent years. Moreover, some companies integrate their environmental and social reporting into their business reports and publish only one integrated report.

While reports addressing single aspects of corporate sustainability can be of certain use, stand-alone sustainability reports and fully integrated corporate reports have received particular attention, especially among large companies. In certain parts of the corporate sector the number of stand-alone sustainability reports nowadays exceeds those of environmental reports. Likewise with environmental statements, there is a trend towards more integrated reporting (BMU 2007). Important drivers for the integration of environmental and social information in financial and annual reports are the increasing interest of investors and analysts as well as regulatory requirements (Hesse 2010; UNEP et al. 2010). The conflation of corporate governance, financial and sustainability reporting has recently been reinforced by the establishment of the International Integrated Reporting Committee (www.theiirc.org). Besides, stakeholder-specific sustainability reporting has been made easier through the technological developments of the Internet. The many advantages of the Internet have made it possible to publish short stand-alone reports that are linked to more in-depth information provided on the corporate websites. Whereas innovation in reporting formats is welcome, and can increase transparency and stakeholder involvement, the continuous experimentation and change of reporting contents and formats by companies, sometimes from year to year, can hamper its comprehensibility and comparability. Further problems and challenges of sustainability reporting are described next.

Specific Challenges in Sustainability Reporting

Corporate sustainability reporting is characterized by several specific challenges:
- Agreement over the terms *sustainable development* or *corporate sustainability* is usually rather difficult *as their meaning is context specific. In practice,* the meaning of corporate sustainability is often not made explicit or constructed as the usual win-win rhetoric (Laine 2005). Moreover, the terminology applied to non- or

extra-financial reporting initiatives varies greatly and changes fairly fast and often. Owen and O'Dwyer (2008) see a tendency for 'corporate responsibility' to displace 'sustainability' and 'social and environmental'. More recently, 'environmental, social and governance' (ESG) seems to have become the term of choice (UNEP et al. 2010). In consequence, management is challenged to define and communicate its understanding of corporate sustainability and establish an approach to identify contextual priorities of non- or extra-financial reporting.

- The complexity of corporate sustainability as a set of interrelated goals leads to problems for management in operationalisation, measurement and communication. It is often difficult to identify and analyse sustainability issues as this requires a change in current and traditional thinking and perceptions. Moreover, little is known about the implementation of accounting and information management systems that would provide a comprehensive basis to identify and report on sustainability issues as well as about how to link strategic analysis and management with information management, corporate accounting and sustainability reporting.
- Sustainability reporting often focuses on *performance* measurement and lacks responsiveness to stakeholder concerns by leaving out *impacts* of corporate activities that are material to key stakeholder groups. The question of course arises as to whether the potential for corporate sustainability reporting to demonstrate accountability for material social and environmental impacts is somewhat limited by nature. As Gray (2006) states, "Precise, reliable statements of organisations' sustainability are oxymorous. Sustainability is a planetary, perhaps regional, certainly spatial concept and its application at the organisational level is difficult at best." Interesting attempts to integrate various societal actors within reporting have recently been made in Italy with the concept of the "bilancio sociale territoriale" – expressing a commitment to report on impacts more broadly.
- There is a two-fold *information asymmetry* between the company and its stakeholders. Stakeholders can often access information about the sustainability of a company only with difficulty and its acquisition can involve very high costs in both time and money (Schaltegger 1997). This information asymmetry tends to create a *climate of low credibility*. On the contrary, companies may not always have sufficient *knowledge about the information needs of stakeholders*. As a result sustainability reports do not always meet stakeholders' information needs and often only a small part of the desired readership is actually contacted (ECC 2003). Although the latter is a common fate of communication it is interesting to note that to date only a limited number of systematic and comprehensive studies has been conducted on stakeholders' reception of and attitudes towards sustainability disclosure practice (Tilt 2007; Owen and O'Dwyer 2008).
- Sustainability reports have often been criticised for being non-specific, aiming at a diffuse and excessively wide group of potential readers. *This lack of target group orientation* creates a risk of *information overload*. The term 'carpet bombing syndrome' (SustainAbility and UNEP 2002) illustrates the fact that some companies have 'flooded' their readers with increasingly extensive sustainability reports – noted by some, but in practice mostly read by only a few.

- The comparability of ecological and social performance information published in sustainability reports is often limited. A *generally accepted standard* about what information should be disclosed and in what format is missing. Moreover, the procedures and practices of data collection and information management can vary over time or between companies.
- Finally, sustainability reporting still remains more common with large and publicly listed companies. Since small and medium sized enterprises constitute a large part of the economy and (in total) account for much of the social and environmental effects of business, a particular challenge seems to lie in encouraging them to engage with forms of sustainability disclosure and reporting, e.g. by emphasising the benefits and keeping the costs of sustainability reporting low.

These challenges complicate the development of confidence and credibility in communication processes within companies, as well as between enterprises and their stakeholders. The next section provides an overview of current developments relevant to overcoming the problems discussed above.

Current Developments

Guidelines and Standards

Various national and international bodies have published guidelines, standards, regulations or sets of criteria aiming at helping to *harmonise sustainability reporting* and providing *guidance for management* in the reporting process. A *guideline* is a non-binding guidance document published by a governmental or non-governmental organisation and often based on practical experiences. Guidelines often precede standards or regulations. Reporting *regulations* are issued by associations and ministries and have a binding character (see next section). They can be based on *standards* which, in turn, are developed by standardisation organisations and are a common basis for certification procedures.

The G3 guidelines of the GRI (2006) are certainly the *most generally accepted and universally applied sustainability reporting framework* and considered to be a *de-facto standard*. Other bodies which have developed *international guidelines and standards* are, for example, the World Business Council for Sustainable Development (WBCSD 2002) and the International Organization for Standardization (ISO 2006). Comprehensive overviews of international guidelines and standards are provided by Adams and Narayanan (2007) and Leipziger (2010). Voluntary guidelines and standards at the *regional or national level* are issued by many *governments* as well as *non-governmental bodies* such as industry associations or other private institutions. In Europe there are also guidelines specifically addressing environmental statements.

The various guidance documents can differ in the particular aspects of sustainability, the sector and the size of companies that they address as well as in the extent to which they focus on reporting principles and report content. Sector-specific guidelines for the production of environmental or sustainability reports that aim at covering all main activities in a specific sector exist at international (sector supplements of G3 guidelines), regional and national levels. Some guidance documents specifically address SMEs. For example, the GRI (2004) has developed a 'beginner's guide' which is particularly targeted at small and medium-sized enterprises.

Overall, a growing body of international and national guidance documents for sustainability reporting has evolved in recent years. In one of the most recent reviews of approaches to enhance sustainability reporting in 30 selected countries (UNEP et al. 2010) a total of 50 standards, codes and guidelines with some form of voluntary sustainability-related reporting guidance have been identified. The study does not provide a comprehensive list but it indicates an *increasingly diverse and mature international framework* for sustainability reporting and *calls for closer collaboration* between standard or guideline setting bodies. The GRI appears to play a key role in achieving a convergence of the various approaches. The strategic alliance of the GRI and the UN Global Compact (UNGC) can be seen as one of the first initiatives to reduce the complexity of reporting practices. Since 2010, the UNGC encourages its participants through its reporting policy to use the G3 guidelines when demonstrating progress towards attainment of the ten principles of the UNGC within their annual Communication on Progress (COP) reports. Easier understanding of the (increasingly confusing) multitude of reporting schemes is facilitated by various types of reporting linkages between the GRI and bodies such as the UN Principles for Responsible Investment Initiative, the Organisation for Economic Co-operation and Development, and the ISO (UNEP et al. 2010). It is generally assumed that enhancing the *convergence* between the numerous reporting schemes will *strengthen their adoption and implementation*. Adams and Narayanan (2007) however stress that guidance documents differ in terms of the extent to which they are concerned with the interest of business and the views and needs of a broad range of stakeholders. There remains a *tension* between using reporting guidelines as a legitimating exercise (to report the minimum required in such guidelines) and demonstrating accountability for the views and needs of a broad range of stakeholders. Given that some guidelines focus on the needs of business and prescribe report content at the expense of concern with processes of stakeholder engagement, they conclude that "[…] without mandatory reporting guidelines focusing on processes of reporting and governance structures, some companies will continue to produce reports which leave out impacts which are material to key stakeholder groups" (Adams and Narayanan 2007: 83). Indeed, as discussed below, many governments have begun to determine mandatory reporting requirements and setting regulatory frameworks for reporting.

Regulations

The disclosure of sustainability information has become the subject of a growing body of regulations (UNEP et al. 2010). In Europe, the implementation of the EU Accounts Modernisation Directive, a reform regulating the balance sheet (EU 2003), has forced shareholder companies to include non-financial performance indicators, specifically also environmental and labour-related indicators, in the prognosis reports included in their annual reports. Mandatory regulations with an obligation to publish sustainability reports (e.g. Denmark, Sweden) or integrated annual reports (e.g. France) provide further evidence for the heightened attention given to the regulation of sustainability reporting in Europe (for a review of such legislation showing that for a number of European countries it is not a recent phenomenon, see IIIEE 2002). Also other parts of the world have been experiencing a move towards mandatory sustainability reporting in recent years (e.g. South Africa). According to the UNEP et al. (2010) study, out of 142 country standards identified in 30 selected countries that are related to some form of sustainability reporting, approximately two thirds can be classified as mandatory. These regulations however are mostly limited to companies of a certain size, state-owned or listed companies, or companies that are significant emitters.

For many years, there has been a lively debate about the role governments should play in sustainability reporting. Some researchers have called for governments to put in place at least a minimal regulatory framework in order to overcome the incompleteness of voluntary non-financial reporting and the reluctance of a vast majority of companies to making any kind of sustainability disclosure. This would prevent companies from conveying a misleading view of their activities and seeking to manage public impressions in their own interest through the provision of false information (Adams and Narayanan 2007; Gray 2006). While the proponents of mandatory reporting note that "waiting for voluntary reporting standards or the merits of peer pressure to raise the bar for everyone is overly optimistic and naive" (Buhr 2007) and no longer an adequate option given the importance of corporate impacts on the environment and society as a whole, sceptics have often questioned that regulations (alone) can have a significant impact on both corporate accountability and the quality of sustainability information published in reports (Owen et al. 1997; Schaltegger 1997) or stressed that command and control regulation may not only be costly but also stifle innovation (Buhr 2007). Rather than advancing voluntary reporting to the detriment of the regulation of disclosures, concern has been raised about a too "simplistic view, according to which the regulation of environmental reporting would prevent all the shortcomings of voluntary environmental disclosures" (Larrinaga at al. 2002: 737). Using the theoretical distinction made by Owen et al. (1997) between administrative and institutional reform, Larrinaga et al. (2002) conclude in their analysis of the Spanish environmental disclosure standard that at a minimum more participation in the form of discursive dialogue is needed in the development of regulation and the effective enforcement of legislation (see also Owen and O'Dwyer 2008). Using an information econometrics perspective, Schaltegger (1997) argues that reporting regulations do not necessarily improve the information situation for stakeholders as companies with passive or indifferent environmental strategies may focus on reducing their reporting costs to meet the regulatory requirements by neglecting the quality of information in their information management procedures. This can lead to an adverse selection in reports whereby bad information quality drives out good information quality (Schaltegger 1997).

In recent years 'smart regulations' have become more prominent that consider voluntary and mandatory reporting to be a spectrum rather than conflicting positions and that hold regulations (by a variety of institutional bodies such as governments, accounting regulators or securities regulators) unfold their highest potential when designed as complementary approaches that enhance sharing relevant information (Buhr 2007). Smart regulations make use of a variety of forms for regulating sustainability reporting (e.g. mandatory regulations, incentives, endorsements, and agreements/non-enforceable contracts with regulators; Gunningham and Grabosky 1998). One of the most recent examples is Denmark, which introduced new legislation requiring companies to disclose CSR information or to explain why they could not (report-or-explain approach) and, by exempting companies from reporting that have acceded to the UNGC and issue COPs, promotes a stronger connection of the various initiatives while counteracting the proliferation of national standards.

An area that has yet not been addressed by many government regulators is the assurance of sustainability reports. With a few exceptions (e.g. Swedish state-owned companies since 2007), and in contrast to financial reporting, companies are not required to subject their sustainability reports to external assurance. Developments in assurance and auditing of sustainability reports are described next.

Assurance, Assessment, and Auditing

Verification of published information is common and is often required for financial reports but in recent years sustainability reports have received greater attention as well (Kolk and Perego 2010; KPMG 2008). The aim of *assuring, assessing and auditing information* disclosed in corporate reports is to help improve their credibility. The G3 guidelines of the GRI (2006) recommend the assurance of sustainability reports and make it compulsory for those companies aiming at achieving the + level of compliance. A survey of the 100 largest companies from 22 countries by KPMG (2008) showed that 45% issue a separate non- or extra-financial report; and of these, 39% had their reports verified. Assurance of sustainability reports has however not achieved equal prominence in all countries. There are considerable differences in the number of assured sustainability reports, with France (73%) and Spain (70%) leading the way and countries such as Romania (4%), United States (14%) and Canada (19%) providing only low levels of sustainability assurance. This is probably largely due to the overwhelmingly voluntary nature of assurance of sustainability reports and the lack of a generally accepted standard in this field. The Federation des Experts Comptables Europeens (FEE) and the national accountancy bodies have been particularly proactive in putting pressure on the International Auditing and Assurance Standards Board to develop an international standard or guidance document for assurance of sustainability reports (UNEP et al. 2010).

The current assurance landscape is characterised by a wide variety of national and international initiatives. The UNEP et al. update (2010) on trends in approaches to sustainability reporting identifies a total of *14 assurance standards*. The two *international standards* increasingly applied for the assurance of sustainability reports are the International Standard on Assurance Engagements 3000 of the International Federation of Accountants and the Assurance Standard AA1000 of the Institute of Social and Ethical Accountability. At a *national* level, standards include for example the German auditing standard 'Generally accepted assurance standards for the audit or review of sustainability reports' and the Dutch standard RL 3410 on assurance engagements relating to sustainability reports. In other European countries such as France, Spain and Sweden (and outside Europe in countries such as Australia, China, and Japan) similar country-specific standards have been drafted or published by national accountancy bodies.

The *contribution assurance practice is making towards enhancing the level of environmental and social accountability* to the stakeholders of companies has been called into question by a number of studies. O'Dwyer and Owen's (2008) review of the growing body of work in this field reveals the *limited contribution to promoting greater transparency and true accountability to the stakeholder* due to ambiguities and inconsistencies in current approaches to sustainability assurance practice (such as independence and degree of thoroughness of audits, managerial control over the whole assurance process, and great uncertainties in understanding the relationship between actual performance and aspects covered by the assurance process). They point out that unless institutional reform empowers stakeholders, i.e. transfers some degree of power over the assurance process from companies to stakeholders and institutes more participatory forms of corporate governance, it is difficult to avoid concluding that little or no value has been added to the assurance process. Whether the developments in mandatory sustainability reporting will further more sophisticated standards for the assurance of sustainability reports (UNEP et al. 2010) remains to be seen.

A particular challenge for the assurance of sustainability reporting lies in web content as no standard for assessing and certifying web content exists and information on the Internet can be changed quite easily. However, the limitations of printed reports as discussed in the next section has encouraged companies to turn to the more expansive possibilities provided by the Internet.

Internet Support

When reviewing studies of online sustainability reporting over the last 10 years or so, it becomes obvious that there has been an overall increase in using the Internet for sustainability reporting (e.g. Herzig and Godemann 2010; Morhardt 2010). Greater use of this new communication approach is often attributed to its advantages in *providing more sustainability information* and *increasing information accessibility and comprehensibility* (Adams and Frost 2006; Herzig and Godemann 2010). With the media-specific linking possibilities and the use of the HTML format, reporting is for example no longer limited by the number of printed pages. A large quantity of information, including historical company information and links to other information sources related to the company or to other organisations can be offered online without necessarily creating a 'carpet bomb' for the reader.

The Internet allows a company to present an integrated view of different aspects of sustainability and interested stakeholders to select, from a large information data base, that information which is of specific interest to them. Moreover, the Internet offers possibilities such as 24-hour accessibility, addressee-specific information tailoring and distribution, individual access for stakeholders, and the combination of different media elements such as words, figures, images or videos (Isenmann 2005). Finally, *there is a much greater range of communication possibilities through stakeholder engagement and dialogue tools in online sustainability reporting* than in printed sustainability reports (Unerman and Bennett 2004; Unerman 2007). While in printed reports stakeholder dialogue mainly takes place prior to production and results of these stakeholder engagement processes can be documented in the reports, *dialogue-based online relationships can include various forms of dialogue* (mutual asynchronous forms such as mail-to functions or discussion forums as well as mutual synchronous forms such as chats, audio or video-conferencing).

However, there are also several *disadvantages to using the Internet*. Information on websites can be changed and the assurance of web content is difficult. Traditional printed reports are thus often said to enjoy a higher degree of credibility among users than online reports. Because some stakeholders tend to be excluded from the Internet or hindered in their use of it, and as stakeholders and reading situations may still favour a printed report, their combined use is usually recommended as the primary way of communication. On the other hand, considerations to encourage a wider application of EMAS by reducing the costs of publishing environmental statements has raised a debate about the necessity of printed reports. Also, in practice some companies have abandoned printed reports and now concentrate on Internet-based sustainability reporting (e.g. Adidas or E.ON UK).

Overall, most of the studies mentioned above reveal that to date many *companies do not make use of the Internet's full potential for sustainability reporting*. Research into trends in the use of Internet-supported sustainability reporting for German DAX30 companies over a 3-year period provides a graphic illustration of the increased consideration of Internet features for reducing information costs for companies and stakeholders but not for enhancing corporate value through more intensive and credible dialogue (Herzig and Godemann 2010). There remains potential for improvement, particularly in the use of tools for stakeholder dialogue, in the introduction of, for example, multimedia elements and in the use of other Internet technologies to improve the dissemination of past and present information (e.g. through archives, website updates, news, and newsletters). Still, there is a growing body of evidence for using the Internet to provide information to various stakeholders. Among the stakeholders who use this information are ranking and rating organisations.

Rankings and Ratings

For sustainability research and rating organisations, sustainability reporting has become an *important source of information*. In addition to data from questionnaire responses from management, these organisations use publicly available company documentation such as sustainability reports and company websites related to sustainability issues.

Furthermore, sustainability reports have themselves increasingly become a subject of *rankings and reporting competitions* reflecting the expectations of stakeholders and possibly involving recommendations for future improvement. By this means, rankings aim to advance the field of sustainability reporting and, to some extent, also contribute to a certain degree of standardisation.

The first rankings of environmental reports were conducted in European countries in the middle of the 1990s. Since 1996, the '*European Sustainability Reporting Awards*' (ESRA) has annually awarded the best external environmental and sustainability reports of private as well as public organisations across Europe. The participants in this European competition (formerly '*European Environmental Reporting Awards*', EERA) were accountancy bodies from 15 European countries, each of which conduct separate national reporting schemes and submit the national winning reports to the European Sustainability Reporting Awards. Separate awards were given to large companies and to small and medium-sized enterprises respectively. In 2006, the European Awards scheme was stopped and ESRA renamed into '*European Sustainability Reporting Association*' whose purpose is to share European reporting developments based on annual reports from each participating country (see www.sustainabilityreporting.eu). Nevertheless, rankings of sustainability reports represent a continuing element of research on sustainability reporting – in Europe and internationally (for an overview see Morhardt 2010).

Outside-In and Inside-Out Perspective

Many of the guidelines and ratings introduced above provide criteria that are used by corporate reporting providers to improve their sustainability reporting. In the extreme, the orientation towards stakeholder requirements can be seen as an '*outside-in*' *approach* towards designing the reporting, accounting and communication process (see Schaltegger and Wagner 2006; Herzig and Schaltegger 2006). With this approach the company analyses stakeholder dialogues and screens the information demand of stakeholders to define its key indicators for reporting and the underlying accounting and data collection processes. The aim is to fulfil external information requests and to provide the information that stakeholders are interested in receiving (e.g. meeting the demands of rating agencies and excelling in external benchmarking schemes and reporting awards). This approach contrasts with the strategic '*inside-out*' *approach* of sustainability performance measurement, management and reporting in which managers first analyse the company's main sustainability weaknesses, then design problem solutions, implement them, establish a measurement and indicator system, and set up a sustainability accounting and data monitoring system in order finally to report the actual situation, the achievements and the goals for future improvements.

The outside-in approach to sustainability reporting has its *strengths and weaknesses*. It is geared towards stakeholder perceptions, media attention and improving rating results, and prevents management from sub-optimising reporting in relation to stakeholder preferences and reactions. Although the outside-in approach is by its nature more reactive and adaptive than the inside-out approach, the latter may tend to neglect some issues that are considered important by some relevant stakeholders. Only a sufficient engagement with stakeholders and consideration of external criteria schemes and requirements can ensure that the company acts in accordance with society's perceptions and goals.

Nevertheless, taken to its extreme, the outside-in approach implies a risk that information is generated and reported without sufficient critical reflection on the themes and corporate activities that are actually relevant for successful sustainable business development. External stakeholders usually do not have sufficient knowledge about production processes, product formulae, etc. to judge the main corporate weaknesses, and to know which changes are necessary on the journey towards sustainable organisation and business development. However, this does not mean that general criteria catalogues, ratings and competitions are pointless, since they constitute important drivers of sustainability reporting and often also of sustainability management. However, with their fairly general character, they have only a limited effect in achieving a substantial improvement in sustainability reporting and corporate sustainability since they cannot cover the necessary details of all issues relevant to a company's sustainable development. Therefore, we suggest embedding *sustainability reporting in a double-path approach that combines the strategic inside-out approach of performance measurement and management with the outside-in approach of adapting to the external requirements* (Fig. 14.2).

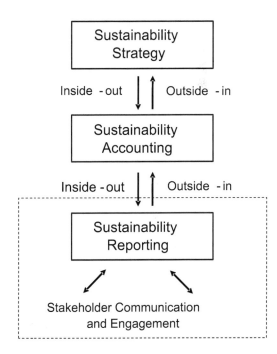

Fig. 14.2 Outside-in and inside-out approach to sustainability reporting

Consequences for Sustainability Communication

Our overview of the history of corporate sustainability reporting and the development of approaches to improve sustainability reporting suggests that this field of sustainability communication is characterised as an *ever-changing and dynamic corporate response to different societal challenges and information needs and the expectations* of various stakeholders. Both *theoretical motivations* and *actual challenges* in sustainability reporting have been discussed. A large range of *approaches to overcome problems* in sustainability reporting has also been considered, often emphasising the *role of external bodies* in defining reporting expectations and requirements. This overview shows that there is an increasingly dense network of national and international reporting standards, codes, guidelines and legislation.

However, bringing organisational performance and reporting in line with external requirements is not sufficient for exploring the full potential contribution of companies to sustainable development. Guidelines, standards, ratings as well as auditing and verification processes may well provide assistance for management in designing corporate sustainability reports, but *systematically linking corporate strategy, information management and reporting activities from an outside-in as well as an inside-out perspective* is an important prerequisite to effective reporting. As long as companies do not explain how they identify, analyse and manage those aspects of their business that are most relevant to making a contribution to sustainable development and corporate strategy, while also responding to external expectations and reporting on all of these issues, sustainability reporting still carries a risk in being seen as green washing or a public relations exercise aiming to solely improve corporate reputation.

To avoid the impression that sustainability reporting is used as a public relations tool to mask companies' actual socially and ecologically unsustainable practices, companies need to *engage with stakeholders through 'true' dialogue* (Bebbington et al. 2007). In this chapter the actual and potential role of the Internet in engaging with stakeholders through online relationships and dialogue, and using the technical features of the Web to learn about stakeholders' views, expectations and information needs was discussed. There is, however, a variety of forms of stakeholder engagement that can create and support 'responsible' stakeholder communication (Crane and Livesey 2003; Unerman 2007).

Finally, the challenges and developments considered above suggest that there is great pressure on *corporate actors who are involved in sustainability reporting*, going beyond important knowledge about external rating and assessment schemes, evaluation criteria and sustainability trends in the media. A *well-managed, interdisciplinary team-based process* seems to be required, one that involves different departments, external stakeholders and possibly communication agencies, as well as *diverse competencies* in identifying the sustainability issues that are most relevant to both the company and society. Likewise, *communicating* these issues in a *comprehensible way* and *integrating sustainability reporting with other sustainability communication media and the company's more general corporate communications concept* appears to be vital if sustainability communication is to move to a higher level.

References

Adams, C., & Frost, G. (2006). Accessibility and functionality of the corporate web site: Implications for sustainability reporting. *Business Strategy and the Environment, 15*, 275–287.

Adams, C., & Narayanan, V. (2007). The 'standardization' of sustainability reporting. In J. Unerman, J. Bebbington, & B. O'Dwyer (Eds.), *Sustainability accounting and accountability* (pp. 70–85). London: Routledge.

Bebbington, J., Brown, J., Frame, B., & Thomson, I. (2007). Theorizing engagement: The potential of a critical dialogic approach. *Accounting, Auditing and Accountability Journal, 20*, 356–381.

Bebbington, J., Larrinaga, C., & Moneva, J. M. (2008). Corporate social reporting and reputation risk management. *Accounting, Auditing and Accountability Journal, 21*, 337–361.

Buhr, N. (2007). Histories of and rationales for sustainability reporting. In J. Unerman, J. Bebbington, & B. O'Dwyer (Eds.), *Sustainability accounting and accountability* (pp. 57–69). London: Routledge.

Bundesumweltministerium (BMU) (German Federal Ministry for the Environment). (2007). *EMAS. Von der Umwelterklärung zum Nachhaltigkeitsbericht (EMAS. From the environmental statement to the sustainability report; only available in German)*. Berlin: BMU.

Crane, A., & Livesey, S. (2003). Are you talking to me? Stakeholder communication and the risks and rewards of dialogue. In J. Andriof, S. Waddock, B. Husted, & S. S. Rhaman (Eds.), *Unfolding stakeholder thinking 2. Relationships, communication, reporting and performance* (pp. 39–52). Sheffield: Greenleaf.

Deegan, C. (2002). The legitimising effect of social and environmental disclosures: A theoretical foundation. *Accounting, Auditing and Accountability Journal, 15*, 282–311.

Dierkes, M., & Antal, B. (1985). The usefulness and use of social reporting information. *Accounting, Organizations and Society, 10*, 29–34.

ECC Kohtes Klewes. (2003). *Global stakeholder report 2003. Shared values? The First World-Wide stakeholder survey on non-financial reporting*. Bonn: ECC Group.

European Union (EU). (2003). Directives 2003/51/EC of the European Parliament and of the Council of 18 June 2003 amending Directives 78/660/EEC, 83/349/EEC, 86/635/EEC and 91/674/EEC on the annual and consolidated accounts of certain types of companies, banks and other financial institutions and insurance undertakings. *Official Journal of the European Union*, 17th July 2003.

Fombrun, C. (1996). *Reputation: Realizing value from the corporate image*. Boston: Harvard Business Press.

Global Reporting Initiative (GRI). (2004). *High 5! Communicating your business success through sustainability reporting. A guide for small and not-so-small businesses*. Amsterdam: GRI.

Global Reporting Initiative (GRI). (2006). *Sustainability reporting guidelines 2006*. Amsterdam: GRI.

Gond, J.-P., & Herrbach, O. (2006). Social reporting as an organisational learning tool? A theoretical framework. *Journal of Business Ethics, 65*, 359–371.

Gray, R. H. (2006). Social, environmental and sustainability reporting and organisational value creation? Whose value? Whose creation? *Accounting, Auditing and Accountability Journal, 19*, 793–819.

Gray, R. H., Kouhy, R., & Lavers, S. (1995). Corporate social and environmental reporting: A review of the literature and a longitudinal study of UK disclosure. *Accounting, Auditing and Accountability Journal, 8*, 47–77.

Gunningham, N., & Grabosky, P. (1998). *Smart regulation: Designing environmental policy*. Oxford: Oxford University Press.

Guthrie, J., & Parker, L. D. (1989). Corporate social reporting: Emerging trends in accountability and theory. *Accounting and Business Research, 19*, 343–352.

Hemmer, E. (1996). Sozialbilanzen. Das Scheitern einer gescheiterten Idee [Social reporting]. The failure of a failed idea. *Arbeitgeber, 23*, 796–800.

Herzig, C., & Godemann, J. (2010). Internet-supported sustainability reporting: Developments in Germany. *Management Research Review, 33*, 1064–1082.

Herzig, S., & Schaltegger, S. (2006). Corporate sustainability reporting: An overview. In S. Schaltegger, M. Bennett, & R. Burritt (Eds.), *Sustainability accounting and reporting* (pp. 301–324). Berlin: Springer.

Hesse, A. (2010). *SD-KPI Standard 2010–2014. Sustainable development key performance indicators: Minimum reporting standard for relevant sustainability information in annual reports/ management commentaries of 68 industries*. Münster: Sustainable Development Management.

International Institute for Industrial Environmental Economics (IIIEE). (2002). *Corporate environmental reporting. Review of policy action in Europe*. Lund: IIIEE.

Isenmann, R. (2005). Corporate sustainability reporting: A case for the internet. In L. Hilty, E. Seifert, & R. Treibert (Eds.), *Information systems for sustainable development* (pp. 164–212). Hershey: Idea Group.

Kolk, A. (2004). A decade of sustainability reporting: Developments and significance. *International Journal for Environmental and Sustainable Development, 3*, 51–64.

Kolk, A., & Perego, P. (2010). Determinants of the adoption of sustainability assurance statements: An international investigation. *Business Strategy and the Environment, 19*, 182–198.

KPMG International. (2008). *KPMG international survey of corporate responsibility reporting 2008*. Amsterdam: KPMG.

Laine, M. (2005). Meanings of the term 'sustainable development' in Finnish corporate disclosures. *Accounting Forum, 29*, 395–413.

Larrinaga, C., Carrasco, F., Correa, C., Llena, F., & Moneva, J. M. (2002). Accountability and accounting regulation: the case of the Spanish environmental disclosure standard. *European Accounting Review, 11*, 723–740.

Leipziger, D. (2010). *The corporate responsibility code book*. Sheffield: Greenleaf.

Morhardt, J. E. (2010). Corporate social responsibility and sustainability reporting on the internet. *Business Strategy and the Environment, 19*, 436–452.

Owen, D., & O'Dwyer, B. (2008). Corporate social responsibility: The reporting and assurance dimension. In A. Crane, A. McWilliams, J. Moon, & D. S. Siegel (Eds.), *The Oxford handbook of corporate social responsibility* (pp. 384–409). New York: Oxford University Press.

Owen, D., Gray, R. H., & Bebbington, J. (1997). Green accounting: Cosmetic irrelevance of radical agenda for change. *Asia-Pacific Journal of Accounting, 4*, 175–198.

Roberts, R. W. (1992). Determinants of corporate social responsibility disclosure: An application of stakeholder theory. *Accounting, Organizations and Society, 17*, 595–612.

Schaltegger, S. (1997). Information costs, quality of information and stakeholder involvement. *Eco-Management and Auditing, 4*(3), 87–97.

Schaltegger, S., & Sturm, S. (1990). Ökologische Rationalität [Ecological rationality]. *Die Unternehmung, 44*, 273–290.

Schaltegger, S., & Wagner, M. (2006). Integrative management of sustainability performance, measurement and reporting. *International Journal of Accounting, Auditing and Performance Evaluation, 3*, 1–19.

Schmidheiny, S. (1992). *Changing course: A global business perspective on development and the environment*. Cambridge: MIT Press.

Spence, C., & Gray, R. H. (2007). *Social and environmental reporting and the business case*. Research Report 98. London: Association of Chartered Certified Accountants.

SustainAbility, & UNEP. (2002). *Trust us. The global reporters. 2002 survey of corporate sustainability reporting*. London: SustainAbility.

Tilt, C. A. (2007). External stakeholders' perspectives on sustainability reporting. In J. Unerman, J. Bebbington, & B. O'Dwyer (Eds.), *Sustainability accounting and accountability* (pp. 104–126). London: Routledge.

Tinker, T., Lehman, C., & Neimark, M. (1991). Falling down the hole in the middle of the road: Political quietism in corporate social reporting. *Accounting, Auditing and Accountability Journal, 4*, 28–54.

Ullmann, A. A. (1985). Data in search of a theory: A critical examination of the relationships among social performance, social disclosure and economic performance of U.S. firms. *Academy of Management Review, 10*, 540–557.

Unerman, J. (2007). Stakeholder engagement and dialogue. In J. Unerman, J. Bebbington, & B. O'Dwyer (Eds.), *Sustainability accounting and accountability* (pp. 86–103). London: Routledge.

Unerman, J., & Bennett, M. (2004). Increased stakeholder dialogue and the internet: Towards greater corporate accountability or reinforcing capitalist hegemony? *Accounting, Organizations and Society, 29*, 685–707.

United Nations Environment Programme (UNEP), KPMG Advisory N.V., Global Reporting Initiative (GRI) & Unit for Corporate Governance in Africa. (2010). *Carrots and sticks – promoting transparency and sustainability. An update on trends in voluntary and mandatory approaches to sustainability reporting.* UNEP, KPMG, GRI & Unit for Corporate Governance in Africa.

World Business Council for Sustainable Development (WBCSD). (2002). *Sustainable development reporting. Striking the balance.* Geneva: WBCSD.

Chapter 15
Computer Support for Cooperative Sustainability Communication

Andreas Möller

Abstract The following chapter explores computers and communication in organizations from the language-action perspective (LAP). This helps clarify the role of email, instant messaging and online social networks in organizations. But the most important result is that two different forms of coordination of action can be distinguished – communication-based and delinguistified – together with possible transitions between the two. A two-phase approach to sustainable development in organizations is presented. Based on Habermas's theory of communicative action, it is argued that sustainability communication is necessary to overcome traditional generalized actions in conflict with sustainable development. This also requires appropriate computer support.

Keywords Computer • Language-action perspective • Theory of communicative action • Computer support • Sustainability communication

Traditional Relevance of Computers in Organisations

Before discussing the relationship between computers and communication, it is helpful to ask how computers are used in organisations. This seems to be a simple question. Computers are information systems and support rational decision-making. Information instruments are called management information systems (MIS) or decision support systems (DSS) (Keen and Scott-Morton 1978; Orman 1984; Gerson et al. 1992). DSS cover a wide range of functions; "they might simply provide summaries of data; they might forecast future developments in the context of present circumstances or they might simulate the future after some postulated action has

---

A. Möller (✉)
Institute for Environmental and Sustainability Communication,
Leuphana University Lüneburg, Germany
e-mail: moeller@uni.leuphana.de

been taken; they might take account of uncertainties; and they might help the decision makers explore their own perceptions and values" (French and Turoff 2007: 39). All these functions are based on an idea of what people do in organisations: "People process information and make decisions (…); they carry out functional roles, using collections of materials, according to stable rules (…); people create and maintain a structure of authority (…); people negotiate and promote competing interests (…); people enter into personal relationships (…)" (Winograd 1986: 204). Computer systems are used to represent all relevant states and processes within an organization so that decision makers can process these data. The hope is that the software system becomes a 'second world', at least with regard to the organization. Some concepts try to cover supply chains (information instruments for supply chain management) or ecological product life cycles (life cycle assessment). In the perspective of sustainability communication, computers provide information about states and processes within a part of the world. Austin calls such informative statements "constative utterances" (Austin 1971: 13).

Software systems can represent organisations in a new way because computers as machines process symbols and do not – like other machines – process materials and energy. Their purpose is the effective transformation of symbols. This is in line with an understanding of language "as a system of symbols that are composed into patterns that stand for things in the world" (Winograd and Flores 1986: 17). Winograd and Flores call this the concept of correspondence. "(1) Sentences say things about the world, and can be either true or false; (2) what a sentence says about the world is a function of the words it contains and the structures into which these are combined; (3) the content words of a sentence (such as its nouns, verbs and adjectives) can be taken as denoting (in the world) objects, properties, relationships, or sets of these" (Winograd and Flores 1986: 17). Computers 'speak' these languages and process sequences of symbols. Methods for means-end analyses (like cost accounting), discrete event simulation etc. define the grammar of such a language. Computer-based simulation tools use symbol-based immaterial representations to derive step-by-step future states of a system. Because humans understand the language too, they can draw conclusions from the calculated states. For instance, Jay Forrester used a special formal language (System Dynamics or Dynamo, the programming language of System Dynamics) to construct the world models that were the primary data source of the report 'Limits to Growth' (Meadows et al. 1972). Such world models say something about the development of important states of the whole world in the future.

With regard to corporate environmental protection and sustainable development, so-called environmental management information systems are being developed. Environmental management information systems are defined as "organizational-technical systems for systematically obtaining, processing and making environmentally relevant information available in companies. Above all these systems aid in determining the environmental damage caused by companies and designing support measures to avoid and reduce it" (Page and Rautenstrauch 2001: 5). A basic concept is life cycle assessment (LCA). ISO 14040 defines LCA as a "compilation and evaluation of the inputs, outputs and potential environmental impacts of a product system throughout its life cycle" (ISO 14040; Guinée 2002: 5). LCA does not comprise all flows of a material and energy flow system, e.g. a company or supply chain. The relevant flows must be related to a product or service (Consoli et al. 1993; Berlin and Uhlin 2004; Frankl and Rubik 2000). This specifies the intended application context of life cycle assessment. It is designed as a decision support instrument. "A decision-maker uses LCA for generating information on the environmental implications of products. For this purpose a model is set up covering the material and energy flows attributed to a product and their evaluation in view of their environmental impact" (Werner 2005: 5). This perspective results in key architectural decisions of computer-based support systems. EMIS can be characterized as special decision support systems.

Communication is defined as the last step of decision-making, the communication of decisions and results. When external stakeholders of an organization are involved, this kind of communication is called reporting. For example, corporate sustainability management provides "stakeholders with information about sustainability-relevant issues and how the company is dealing with them... An essential goal in informing key stakeholder groups about non-financial issues is to secure the legitimation of corporate activities and the supply of important resources" (Herzig and Schaltegger 2006: 301). Today, reporting is computer-based. This allows target-group tailored reports (Marx Gómez and Isenmann 2004) and interactive reporting (Isenmann and Kim 2006). After all, the role of communication and language is based on the equivalency of management and decision-making, with manager and decision maker being synonymous. Communication is required because other decision makers (stakeholders) need information about the decisions as data input for their own decisions. They need statements about the states and processes relevant for their decisions – and environmental performance is treated more and more as relevant to this decision-making.

Basics of Computer-Supported Communication

Is data exchange between decision makers the only modus of communication in organisations? And is decision-making the only link between communication and action? If this was true all conversation would be characterized as statements about past, current and future situations. Austin, however, emphasized that not all utterances are statements (Austin 1962). He analyzed the relationship between different types of utterances and action. One important distinction is between constative utterances and performance utterances (or performatives). "The constative utterance, under the name, so dear to philosophers, of statement, has the property of being true or false. The performance utterance, by contrast, can never be either: it has its own special job, it is used to perform an action" (Austin 1971: 13). Performatives constitute acts like promising, advising or naming. This theory is called speech act theory. It starts "with the assumption that the minimal unit of human communication is not a sentence or other expression, but rather the

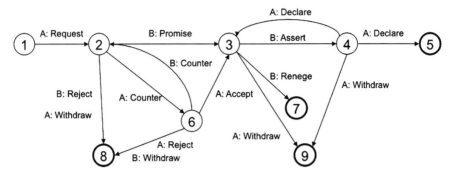

Fig. 15.1 Typical conversation

performance of certain kinds of acts, like making statements, asking questions, giving orders, describing, explaining, apologizing, thanking, congratulating etc." (Searle et al. 1980, p. vii). These acts are called illocutionary acts.

Searle formalized different speech acts (Searle 1969) and constituted five classes of illocutionary acts: assertives, directives (requests), commissives (promises), declarations and expressives. From a computer science viewpoint, such a classification is an important step, because by using these classes it is possible to formalize conversations as a relationship between actors; conversations are processes of these illocutionary acts. A short process is for example 'Request (a directive of actor A) – Commit (a commissive of actor B)' or 'Request (of A) – Counter-offer (a directive of B) – Accept (a commissive of A)'. By folding conversations on the basis of Searle's categories (conversations are processes in time and therefore without loops), typical patterns of conversation can be identified (Fig. 15.1). If we represent each speech act as a symbol and a conversation as a sequence of these symbols, diagrams like Fig. 15.1 define the grammar of a language. Winograd (1986) compares a conversation with a 'dance' – and the term dance is used in both senses of the word, as a 'process' and as a 'system' (like waltz or latin dances).

Computers can be used as conversation support systems. Their purpose is to support effective conversation. Email is a very simple conversation support system. Even if email can be treated as a task management tool supporting project management, task delegation, information handling, scheduling, planning and social communication (Dabbish et al. 2005), email systems normally do not identify patterns of conversation but instead allow users to carry on a conversation. Users can answer an email or they can forward it. Some email clients visualize conversations. If an email is selected, all other emails of the respective conversation are highlighted. However, conversation support of today's email systems is quite poor and technical protocols like SMTP (Simple Mail Transfer Protocol) do not support classes of speech acts. Using subject fields and special character strings like 're' or 'fwd' are a less-than-ideal solution.

Instant messaging (IM) programs facilitate synchronous one-to-one communication between users in a 'buddy list', 'friends list' or in a 'chatroom'. IM programs visualize conversations directly. Message archives consist of three panes: in a pane on the left side all friends, on a top right pane the conversations with a selected friend and on a bottom right pane a selected conversation. But instant messaging does not support speech acts or other classes of utterances. Even if instant messaging is a popular medium for both social and work-related communication (Avrahami and Hudson 2006), effective communication is today not a purpose of instant messaging programs. However, because of its group-oriented functionality (distributed cooperative work, real-time communication, planning social events, socializing) instant messaging supports teams in the workplace. Handel and Herbsleb have analyzed the content of chat and categorized chat content of instant messaging at workplaces. They found that 69% of conversations relates to specific work tasks (Handel and Herbsleb 2002). Other reasons for chatting are negotiating availability (13%), greeting (7%), humour (5%) and non-work (3%). Although work-related content dominates instant messaging at workplaces, data exchange between decision makers does not play an important role. This was a result of sub-classifying 'work'. Handel and Herbsleb write: "We dropped 'walkthrough', 'goal', 'digression' and 'clarification' since we never observed them within the 'work' portion of our protocol" (2002: 6). The most important subcategories were technical work, project management and meeting management. Finally, Handel and Herbsleb pointed out that "chat was used overwhelmingly for work discussions or for articulation work to coordinate projects and meetings, and to negotiate availability" (2002: 8). After all, empirical analyses show that today communication support systems play a critical role in enhancing effective communication in organisations. Moreover, the fact that empirical analyses are required to understand email and instant messaging in organisations emphasizes the high flexibility of these support systems.

These findings are in line with Winograd and Flores's understanding of management beyond decision-making. "In understanding management as taking care of articulation and activation of a network of commitments, produced primarily through promises and requests, we cover many managerial activities. Nevertheless, we also need to incorporate the most essential responsibilities of managers: to be open, to listen, and to be an authority regarding what activities and commitments the network will deal with. These can be characterized as participation in 'conversations for possibilities' that open new background for the conversation for action" (1986: 151). The result that management is more than decision-making is important with respect to corporate sustainability. What are the relationships to sustainability in organisations? And why is this type of conversation necessary in organisations?

Traditional Generalized Action Orientations and the Role of Communication

Even though the challenges of sustainable development have produced successful new scientific communities, the success of these approaches in companies is quite limited. Many companies are discussing concepts of corporate sustainability.

However, cost accounting, cost cutting and labour efficiency are much more important in everyday management in companies than for instance carbon footprints. Concepts like life cycle assessment are welcome if they help to increase economic efficiency. Companies want to identify win-win situations. Contrary to expectations of life cycle management (Remmen et al. 2007), life cycle assessment has more of a supportive function. The question is why.

This question is associated with the ever-present question of a company's responses to its problems. It is not a technical question (e.g. useless corporate information systems or data gaps), it is rather a question of organizational culture and cultural change. Schuhmacher (1997) distinguish two dimensions of cultural change: the number of members and the number of domains changing. If cultural change affects only a few members and the number of domains is low, he characterizes this type of change as a 'drift'. If the number of members and domains changing is high, it is a 'transition'. "Transition is described here as change in many significant domains for a majority of the members of the culture. The magnitude here is so great that the identity of the culture is at least questioned, perhaps redefined" (Schuhmacher 1997: 115). Concepts of life cycle assessment and eco-efficiency are not only information instruments. They are also the 'Trojan horses' of new organizational images and metaphors ('metaphorical thinking', Morgan 1986: 16): carbon-neutral companies as industrial ecosystems, sustainable corporations and sustainable supply chains as socio-economic metabolisms etc. Many members of companies are affected and the changes are significant. So in fact the instruments aim at a transition of societies and of corporations (Fischer-Kowalski and Haberl 2007). The next question is how societies and corporations organize socio-economic transitions.

Habermas's concept of society (1985a, b) distinguishes lifeworld and societal subsystems like the economy. In his concept, economic efficiency (as described by Taylor in 1911), cost cutting and value creation are generalized action orientations in the subsystem economy. It is the result of a "social evolution as a second-order process of differentiation: system and lifeworld are differentiated in the sense that the complexity of the one and the rationality of the other grow" (Habermas 1985b: 153). The process is triggered by communicative action. "Action oriented to mutual understanding gains more and more independence from normative contexts. At the same time, even greater demands are made upon this basic medium of everyday language; it gets overloaded in the end and is replaced by delinguistified media" (Habermas 1985b: 155). That is the paradox of communication. It is a co-evolutionary process and "modern societies attain a level of system differentiation at which increasingly autonomous organisations are connected with one another via delinguistified media of communication: these systemic mechanisms – for example, money – steer a social intercourse that has been largely disconnected from norms and values, above all in those subsystems of purposive rational economic and administrative action that, on Weber's diagnosis, have become independent of their moral-political foundations" (Habermas 1985b: 154).

This concept shows that communication can be understood as the last step of decision-making and as a process of data exchange between decision makers. Communication becomes a part of the systemic mechanisms in the societal subsystem economy.

Moreover, Habermas's concept of society helps explain why corporate sustainability is still a challenge for companies. Sustainability is not a generalized action orientation in the economic system today. Sustainability is ignored as long as it is not compatible with conventional generalized action orientations like economic efficiency and profit maximization. The constellations of compatibility are called win-win situations. Some proposed approaches in the field of environmental accounting are based on the idea of being compatible with dominant action orientations, for example environmental cost accounting or eco-efficiency analysis. However, the term efficiency does not cover important aspects of the underlying action orientation. Economic efficiency is intimately connected with short-term value creation whereas ecological efficiency is more of a contribution to the long-term viability of an organization. In the end, eco-efficiency is still incompatible with today's dominant action orientations in our economy.

How could corporate sustainability become a generalized action orientation in organisations? Habermas 's answer is communicative action. Nowadays, computers can play a prominent role as a powerful medium of everyday communication. But before discussing computer support for communicative action, it is important to keep the limits of communication and discourse in mind. This results in a two-phase approach of sustainable development in organisations: (1) in the first step sustainability communication plays an important role. It supports the organization in questioning tradition. Sustainability communication discloses the organization (Spinosa et al. 2001). In a subsequent step, (2) new routines gradually replace old ones. It is a question of sustainability management and the transformation of new generalized action orientations into effective systemic mechanisms.

Two Phases of Sustainable Development in Organisations

As mentioned above the main problem of communicative action is that it increases complexity. Ideally, all action should be based on communication and discourse. It should be consensus-based action. Computer-based communication support systems like email and instant messaging can increase the efficiency of consensus-based action. And computer systems can help to handle increased complexity by designing social infrastructures that make collective activity visible (Erickson et al. 2002), by visualizing online conversations (Donath 2002), by supporting navigation in social networks with the aid of tag clouds (Mesnage and Carman 2009) etc. Computers as a medium and as a tool increase the domain of consensus-based action. It is then easier for organisations to deal with unexpected situations and ill-defined problems. However, the possibilities of computer support are limited. The application of communicative action is not an all-or-none question. The question is how to organize the relationship between communicative action in organisations and delinguistified coordination of action. To clarify the relationship, three different aspects can be distinguished: (1) Corporate communication is not concerned with operational routine action. The domains of communicative action are non-routine activities. The most important software systems in companies like enterprise resource planning systems support routine jobs in corporations such as purchasing, production, warehousing, book-keeping etc. It is an important job of management to identify optimal standards, which Taylor (1911) calls scientific management. Research is still required to explore new standards in a globalized world, for instance product life cycle management or supply chain management. (2) The problem is that delinguistified rules can come into conflict with emerging challenges like sustainable development. By following the rules, organisations are unable to deal with new challenges. Winograd & Flores call such a situation a breakdown. "By this we mean the interrupted moment of our habitual, standard, comfortable 'being in the world'. Breakdowns serve an extremely important cognitive function, revealing to us the nature of our practices and equipment, making them 'present-to-hand' to us (...) Most important, though, is the fundamental role of breakdown in creating the space of what can be said, and the role of language in creating our world" (1986: 77). In other words, it opens the door for communicative action. (3) Unfortunately, this transition to a different form of cooperative action results in higher complexity. To deal with this higher complexity, complexity in other respects must be reduced. So, the ultimate purpose of communicative action is to replace it by new roles for delinguistified coordination of action. The focus of communicative action is not on isolated problems and solutions in individual cases. It is targeted at abstraction, identification of new ways and new mechanisms. With regard to corporate sustainability, the question is about new standards (in the words of Taylor 'new scientific management'), new forms of corporate information systems etc. However, a direct switch from old to new standards is not possible. New standards are based on new insights, new images and new metaphors as they emerge in communication processes. Images like 'carbon-free company' or 'green company' play a prominent role in such a process. They support the introduction of new information instruments like life cycle assessment. New instruments are tested with the aid of a software tool. Members of the organization become gradually familiar with the new approaches. The results of experiments are presented on PowerPoint slides. The slides are available in internet or intranet as pdf files etc. Sankey diagrams or typical radar diagrams, showing the results of life cycle assessment, become good arguments in such a process.

However, these images, software tools and visualizations do not facilitate the replacement of sustainability communication in organisations by new systemic mechanisms. What is needed is something like 'sustainable business process re-engineering', providing images of business process automation. In fact, some of the first decisions in a transition phase are fairly simple new rules such as the purchase of environmentally friendly office equipment and paper, the activation of energy saving functions of personal computers as a contribution to GreenIT, serving organic food in the canteen etc. But such a set of new roles are not the optimal outcomes of a consistent and integrated concept of sustainable routine in organisations.

Important approaches in creating new integrated ways of doing business include business process re-engineering (Hammer and Champy 1993; Hlupic and Robinson 1998) and business process management (Ko 2009). A business process is defined as an ordering of work activities across time and place, with a beginning, an end, and clearly identified inputs and outputs; or in other words, a structure for action (Davenport 1993). So, the focus of business process management is on the design and control of all relevant business processes within an organization. All questions of cooperation are solved and communication is no longer necessary. It is replaced by business process models and business rule specifications. Software applications can support the design process with the aid of a business process modeller and a business rule editor (Costello and Molloy 2004). A business repository stores the results of the modelling. It is the database of a process engine and a rule engine. Both engines control operational process execution.

Floyd refers to business process models as operational forms or autooperational forms. An operation is treated as a special kind of human activity. "Operation can be analytically separated through scientific observation… Operations are rooted in repeated human action of individuals or groups. Fundamental to operations is the separation of description and performance. The description of operations makes well-defined ways of proceeding possible that can be taught, planned, and enforced. The performance of operations is embedded in situated human activity (…) Operational forms result from connecting the descriptions of individual operations in terms of temporal, logical, and causal relations" (2002: 18).

Floyd points out that operational (re-) construction as a basic method associated with computing is a social process. She talks about the concept of operation, the roles and perspectives of observers, levels of software practice and social reflection. So, the identification of routine and the design of operational and autooperational forms is a social activity. Even if business process re-engineering stands for profit maximization, cost cutting and the release of staff, it is also a non-routine activity. Kieser (1996) has identified in concepts like business process re-engineering preservative organizational patterns: (1) the identification of key factors, with regard to sustainability for example climate change and carbon dioxide, (2) an imminent breakdown (as defined above), e.g. on the basis of the Stern Report, the IPCC reports or an international conference, (3) the problem that central action orientations are endangered, e.g. future profit maximization under conditions of climate change, (4) the presentation of new excellence initiatives as paradigms instead of special instructions, (5) members within an organization who adopt the new ideas as pioneers, (6) very simple ideas and principles are combined with ambiguity, yielding easy to understand action orientations but not simple recipes. All these phases and aspects show that these concepts are mainly based on communicative action (Dietz 2006), e.g. now obvious problems with conventional action orientations, offers of new, hopefully successful ways of thinking etc. In other words, concepts like business process re-engineering can serve as examples or paradigms that allow the identification of development formats for transition processes.

There are many different promising approaches for the computer support of sustainable development. With respect to the role of communication in development and transition processes, two orientations can be distinguished: conversation support and decision support. Computer-based conversation support is especially required to question problematic action orientations and routines. Furthermore, computers as a medium should support the process of finding new, more sustainable routines. In this regard, it is not sufficient to provide web-based chat, wiki and forum components. The transformation process can be treated as a design process with different phases and intermediate results. Analyses of concepts like business process re-engineering provide important hints on how to organize these transition processes.

Computer Support for Sustainability Communication

In the last 10 years, email and instant messaging have become important daily tasks. Today it is imperative that everybody checks their email at least once a day. As discussed before, this can be seen as information exchange between decision makers. But computer-based conversation support systems can also take the form of the new online social networks (OSNs) like Facebook and MySpace. These platforms "allow individuals to (1) construct a public or semi-public profile within a bounded system, (2) articulate a list of other users with whom they share a connection, and (3) view and traverse their list of connections and those made by others within the system" (Weigand and Lind 2008: 51). In other words, they provide new infrastructures of the lifeworld including new domains like messaging, applications (apps), profiles, friends or photos, which become hypertext-based new structures, on a technical level in the form of clusters of web pages. State diagrams show how people move within these structures (Schneider et al. 2009). Web platforms become part of the lifeworld and provide new structures and domains (Fig. 15.2). New Web 2.0 services including micro-blogging, social bookmarking and location-based services (Ullrich et al. 2008) provide a new medium of communication. The Internet can be interpreted as a Piercian 'Pragmatic Web' that integrates different levels of communication: the media level of hardware technologies, the syntactic level for formal languages like HTML and XML, the semantic level of meanings and ontologies and the pragmatic level of information needs, expectations, norms and values (Yetim 2007).

The processes within these networks are characterized as 'socializing' and communication for its own sake (Weigand and Lind 2008). Typical subjects of discussion

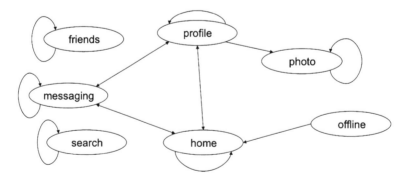

Fig. 15.2 Infrastructure of a web platform representing the user lifeworld

are shared interests, the challenges of daily life, weekend activities, possible face-to-face meetings etc. From an economic perspective the communication processes are trivial and insignificant and would disturb business processes. In the language-action perspective the platforms provide new structures allowing individuals to take up challenges beyond the horizon of systemic mechanisms and traditional action orientations. Spinosa, Flores & Dreyfus characterize this as 'disclosing worlds'. Others characterize this as 'social contagion and the spread of ideas' (Kleinberg 2008).

With regard to sustainable development, the hope is that social networks will contribute to overcoming the conflict between traditional action orientations in societal subsystems (like short-term profit maximization in the economy) and to developing new orientations in line with the ideas of sustainability. OSMs and other Web 2.0 services should open up organisations to these issues. However, there is no guarantee that an organization will decide to become a sustainable organization. This is perhaps the most challenging aspect of all. Members of an organization charged with supporting these processes cannot plan these projects as they can plan and manage traditional projects because the goals and means are not predefined.

There are more fundamental ways in which computer systems can contribute to changing generalized action orientations. Even when computer tools, e.g. for carbon footprinting and life cycle assessment, are developed with decision support in mind, they play an important role in communication processes. Computer tools make problems visible; they help organisations to better understand both problems and possible solutions. In fact, many companies and other organisations make use of life cycle assessment projects and test life cycle assessment tools in order to gain new insights, try to anticipate stricter regulatory standards and explore new competitive opportunities. LCA tools are designed to support these experiments. Computer support in the transition phase has completely different purposes than in phases with stable action orientations. Computers are used as a 'tool' to enhance the capabilities of the users (for more on the tool metaphor see Möller et al. 2006). They provide information on an ad-hoc basis rather than routinely generated data (Burritt et al. 2002). Typical software tools are simulation programs, which for example are based on Forrester's system dynamics approach. Simulation tools require comprehensive formalization efforts because they need all the specifications necessary to calculate the future states of the modelled system. Furthermore, validation steps are required in order to guarantee that the simulation software generates more or less the same process as real processes. However, a special class of software tools does not require these formalization efforts. These tools are used in connection with for example formative scenario analysis (Scholz and Tietje 2002) or the WBGU syndrome concept (WBGU 1996). In fact, these tools are aimed at better understanding ill-defined problems, i.e. problems that are nebulous and unstructured, and new ways of reaching a solution or even solutions that have not yet emerged.

The software instruments serve as a new 'language' in the transformation phase. They are containers of new forms of thinking (e.g. life cycle thinking) and establish new forms of argumentation (e.g. in case of life cycle assessment new typical diagrams and flow charts). Information systems provide data for good arguments, i.e. the data must be presented in a form that fits the communication process.

Good examples for a new 'language' are the Sankey diagrams used to visualize material and energy flows. In the 1990s software tools were developed to support material and energy flow analyses and life cycle assessments. The idea was to present the results in the form of eco-balances (period-oriented input-output balances and life cycle inventories in the product-oriented perspective). In fact, the tools are able to generate these tables with a large number of entries on the input and output side. However, eco-balances are much less popular than expected. It is obviously not possible to use the tables as arguments in communication processes. In fact, the first environmental reports in the 1990s included detailed eco-balances. Reports today present the results in a different way. Surprisingly, an additional visualization instrument to present the results in the form of Sankey diagrams was much more successful. Software development for corporate sustainability has to think about 'languages'. How do methods and tools define new languages? And how can members of an organization understand these new languages? Methods and instruments are still being discussed with regard to correctness (data quality, correctness of methods, system boundaries etc.). The language-action perspective (LAP) sheds light on the question of usability. How can these tools become an important part of conversation support systems? And what are the resulting requirements for the respective software components? Building such software systems is still an unmet challenge (de Moor and Aakhus 2006).

Finally, it is possible to distinguish two different ways in which computer software can support corporate sustainability communication: firstly as a new medium with email, instant messaging, OSNs etc.; and secondly as a support tool for good arguments by providing visualization of non-sustainable structures of value creation. The idea behind this tool is that problematic generalized action orientations cannot be enhanced or replaced directly by new ones.

Conclusions

The most important conclusion from our consideration of images of computer applications in organisations is that computers are regarded traditionally as decision support systems. Computers should support rational decision-making. Communication can be treated as data exchange between decision makers. However, this is accurate only for regular 'business as usual' decisions. But the reality in organisations is quite different; "most, if not all, of a manager's key decisions tend to be fuzzy problems, not well understood by them or the organization, and their personal judgment is essential" (Keen and Scott-Morton 1978: 58). This applies in particular to environmental protection and corporate sustainability, which are not in line with general action orientations in Western economies. Corporate sustainability is obviously not a problem that can be solved by decision makers with the aid of conventional decision support systems. In future, enhanced decision support systems, including components for material flow analysis, life cycle assessment and carbon footprinting, may be able to support organisations that are already sustainable. But today most organisations are not in such a situation. Winograd and Flores conclude that "Instead of talking about 'decisions' or 'problems' we talk of 'situations of irresolution', in which we sense conflict about an answer to the question 'What needs to be done?'" (1986: 147). Here sustainability communication comes into play. Sustainability communication stands for the first step in a two-phase approach to corporate sustainable development.

And software support is required to help effectively answer the question 'What needs to be done?' Here 'decisions' emerge from communication processes within the organization or between the organization and its stakeholders. Speech act theory helps us not only to understand these processes better, it also contains formalization steps that are required for designing effective conversation support systems. The purpose of software systems is not only to support undisturbed communication and not only to find solutions in individual cases. The main purpose of conversation support systems is to identify new – and more sustainable – routines. Taylor would ask for new standards. So, computer support for corporate sustainability should be considered and used for two different reasons: (1) in a phase of transition and in 'situations of irresolution' as an effective communication medium and (2) in everyday routine situations as a decision support system.

References

Austin, J. (1962). *How to do things with words*. Cambridge, MA: Harvard University Press.
Austin, J. (1971). Performative-constative. In J. Searle (Ed.), *The philosophy of language* (pp. 13–22). Oxford, UK: Oxford University Press.
Avrahami, D., & Hudson, S.E. (2006). Communication characteristics of instant messaging: Effects and predictions of interpersonal relationships. *Proceedings of the 2006 20th Anniversary Conference on Computer Supported Cooperative Work*, Banff, Alberta.
Berlin, D., & Uhlin, H.-E. (2004). Opportunity cost principles for life cycle assessment: Toward strategic decision-making in agriculture. *Progress in Industrial Ecology, 1*(1/2/3), 187–202.
Burritt, R., Hahn, T., & Schaltegger, S. (2002). Towards a comprehensive framework for environmental management accounting. *Australian Accounting Review, 12*(2), 93–109.
Consoli, F., Allen, D., Boustead, I., Fava, J. A., Franklin, W. E., Jensen, A. A., de Oude, N., Parrish, R., Perriman, R., Postlethwaite, D., Quay, B., Seguin, J., & Vigon, B. W. (1993). *Guidelines for life-cycle assessment: A code of practice*. Brussels/Washington, DC: Society of Environmental Toxicology and Chemistry (SETAC).
Costello, C., & Molloy, O. (2004). Orchestrating supply chain interactions using emerging process description languages and business rules. *Proceedings of the 6th International Conference on Electronic Commerce* (pp. 21–30). Delft, The Netherlands: ACM.
Dabbish, L. A., Kraut, R. E., Fussell, S., & Kiesler, S. (2005). Understanding email use: Predicting action on a message. *Proceedings of the SIGCHI Conference on Human Factors in Computing Systems* (pp. 691–700). Portland, Oregon, USA: ACM.
Davenport, T. H. (1993). *Process innovation: Rengneering work through information technology*. Cambridge, MA: Harvard Business School Press.
De Moor, A., & Aakhus, M. (2006). Argumentation support: From technologies to tools. *Communications of the ACM, 49*(3), 93–98.
Dietz, J. L. M. (2006). The deep structure of business processes. *Communications of the ACM, 49*(5), 58–64.

Donath, J. (2002). A semantic approach to visualizing online conversations. *Communications of the ACM, 45*(4), 45–49.

Erickson, T., Halverson, C., Kellogg, W. A., Laff, M., & Wolf, T. (2002). Social translucence - designing social infrastructures that make collective activity visible. *Communications of the ACM, 45*(4), 40–44.

Fischer-Kowalski, M., & Haberl, H. (Eds.). (2007). *Socioecological transitions and global change – trajectories of social metabolism and land use.* Cheltenham: Edward Elgar Publishing.

Floyd, C. (2002). Developing and embedding autooperational form. In Y. Dittrich, C. Floyd, & R. Klischewski (Eds.), *Social thinking – software practice* (pp. 5–28). Cambridge, MA: MIT Press.

Frankl, P., & Rubik, F. (2000). *Life cycle assessment in industry and business – adoption patterns, applications and implications.* Berlin/Heidelberg/New York: Springer.

French, S., & Turoff, M. (2007). Decision support systems. *Communications of the ACM, 50*(3), 39–40.

German Advisory Council on Global Change (WBGU). (1996). *World in transition: The research challenge.* Berlin: Springer.

Gerson, M., Chien, I. S., & Raval, V. (1992). Computer assisted decision support systems: Their use in strategic making. *Proceedings of the 1992 ACM SIGCPR Conference on Computer Personnel Research.* New York, NY: ACM.

Guinée, J. B. (Ed.). (2002). *Handbook on life cycle assessment – operational guide to the ISO standards.* Dordrecht/Boston/London: Kluwer Academic.

Habermas, J. (1985a). *The theory of communicative action* (Reason and the rationalization of society 3rd ed., Vol. 1). Boston: Beacon.

Habermas, J. (1985b). *The theory of communicative action* (Lifeworld and system: A critique of functionalist reason 3rd ed., Vol. 2). Boston: Beacon.

Hammer, M., & Champy, J. (1993). *Reengineering the corporation.* New York: Harper Collins Books.

Handel, M., & Herbsleb, J. D. (2002). What is chat doing in the workplace? *Proceedings of the 2002 ACM Conference on Computer Supported Cooperative Work*, Louisiana, USA.

Herzig, C., & Schaltegger, S. (2006). Corporate sustainability reporting. an overview. In S. Schaltegger, M. Bennett, & R. Burrit (Eds.), *Sustainability accounting and reporting* (pp. 301–324). Dordrecht: Springer.

Hlupic, V., & Robinson, S. (1998). Business process modelling and analysis using discrete-event simulation. *Proceedings of the 1998 Winter Simulation Conference.* Washington, DC: IEEE Computer Society Press.

International Standardisation Organisation (1996). Environmental management – Life cycle assessment – Principles and framework – ISO 14040, Paris.

Isenmann, R., & Kim, K.-C. (2006). Interactive sustainability reporting. Developing clear target group tailoring and stimulating stakeholder dialogue. In S. Schaltegger, M. Bennett, & R. Burrit (Eds.), *Sustainability accounting and reporting* (pp. 533–555). Dordrecht: Springer.

Keen, P. G. W., & Scott-Morton, M. S. (1978). *Decision support systems: An organizational perspective.* Reading, MA: Addison Wesley.

Kieser, A. (1996). Moden und Mythen des Organisierens. *Deutsche Betriebswirtschaft, 56*(1), 21–36.

Kleinberg, J. (2008). The convergence of social and technological networks. *Communications of the ACM, 51*(11), 66–72.

Ko, R. K. L. (2009). A computer scientist's introductory guide to business process management (BPM). *Crossroads, 15*(4), 11–18.

Marx Gómez, J., & Isenmann, R. (2004). Editorial: Developments in environmental reporting. *International Journal Environment and Sustainable Development, 3*(1), 1–4.

Meadows, D. H., Meadows, D. L., Randers, J., & Behrens, W. W., III. (1972). *The limits to growth.* New York: Universe Books.

Mesnage, C., & Carman, M. (2009). Tag navigation. *Proceedings of the 2nd International Workshop on Social Software Engineering and Applications* (pp. 29–32). Amsterdam: ACM.

Möller, A., Prox, M., & Viere, T. (2006). Computer support for environmental management accounting. In S. Schaltegger, M. Bennett, & R. Burritt (Eds.), *Sustainability accounting and reporting* (pp. 605–624). Dordrecht: Springer.

Morgan, G. (1986). *Images of organization*. London: Sage Publications.

Orman, L. (1984). A multilevel design architecture for decision support systems. *ACM SIGMIS Database, 15*(3), 3–10.

Page, B., & Rautenstrauch, C. (2001). Environmental informatics - methods, tools and applications in environmental information processing. In C. Rautenstrauch & S. Patig (Eds.), *Environmental information systems in industry and public administration* (pp. 2–13). Hershey/London: Idea Group Publishing.

Remmen, A., Jensen, A. A., & Frydendal, J. (2007). *Life cycle management – a business guide to sustainability*. Paris: UNEP/SETAC Life Cycle Initiative Publications.

Schneider, F. et al. (2009). Understanding online social network usage from a network perspective. *Proceedings of the 9th ACM SIGCOMM Conference on Internet Measurement*, Chicago, IL.

Scholz, R. W., & Tietje, O. (2002). *Embedded case study methods: Integrating quantitative and qualitative knowledge*. Thousand Oaks: Sage.

Schuhmacher, T. (1997). West coast Camelot. The rise and fall of an organizational culture. In S. A. Sackmann (Ed.), *Cultural complexity in organizations. Inherent contrasts and contradictions* (pp. 107–132). Thousand Oaks/London/New Delhi: Sage.

Searle, J. (1969). *Speech acts*. Cambridge, UK: Cambridge University Press.

Searle, J., Kiefer, F., & Bierwisch, M. (1980). Introduction. In J. Searle, F. Kiefer, & M. Bierwisch (Eds.), *Speech act theory and pragmatics* (pp. vii–xii). Dordrecht/Boston/London: D. Reidel Publishing.

Spinosa, Ch, Flores, F., & Dreyfus, H. (2001). *Disclosing new worlds – entrepreneurship, democratic action and the cultivation of solidarity*. Cambridge, MA/London: MIT Press.

Taylor, F. W. (1911). *The principles of scientific management*. New York/London: Harper and Brothers Publications.

Ullrich, C. et al. (2008). Why web 2.0 is good for learning and for research: Principles and prototypes. *Proceedings of the 17th International Conference on World Wide Web*, Beijing, China.

Weigand, H., & Lind, M. (2008). On the pragmatics of network communication. *Proceedings of the 3rd International Conference on the Pragmatic Web: Innovating the Interactive Society*, Uppsala, Sweden.

Werner, F. (2005). *Ambiguities in decision-oriented life cycle inventories*. Berlin/Heidelberg/New York: Springer.

Winograd, T. (1986). A language/action perspective on the design of cooperative work. *Proceedings of the 1986 ACM Conference on Computer-Supported Cooperative work*, Austin, TX.

Winograd, T., & Flores, F. (1986). *Understanding computers and cognition*. Reading, MA: Addison Wesley.

Yetim, F. (2007). Discoursium for cooperative examination of information in the context of the pragmatic web. *Proceedings of the 2nd International Conference on Pragmatic Web*, Tilburg, The Netherlands.

Chapter 16
Participation: Empowerment for Sustainable Development

Harald Heinrichs

Abstract The current discussion about participation and sustainability shows the relevance of participative elements in modern societies for coping with social, ecological and technological complexity. Especially in the 1990s participative processes for cooperative planning and decision-making procedures were developed and tested. Faced with the challenges confronting society as it moves towards a more sustainable development, both the opportunities and the limits of a culture of participation and sustainability will become more and more noticeable.

Keywords Participation • Cultural evolution • Cooperation • Participation methods • Sustainability communication

Participation and Sustainable Development

Since the 1990s two terms have had an impressive career in national and international discourses about the viability of (global) society: participation and sustainability. Since the United Nations Conference on Environment and Development in Rio de Janeiro in 1992, participation and sustainability have become commonplace topics in academic articles, journalistic commentaries and political discussions. In the wake of the financial and economic crisis following 2008, there has been an increase – at least rhetorically – in attention paid to the social, ecological and economic dimensions of the problems facing society, as well as the importance of intra – and intergenerational justice and the role of heterogeneous groups of actors in solving these problems.

H. Heinrichs (✉)
Institute for Environmental and Sustainability Communication, Leuphana University Lueneburg, Lueneburg, Germany
e-mail: harald.heinrichs@uni.leuphana.de

Sustainability and participation have thus become important aspects in a number of current debates, whether the reform of the financial, health or social systems, the securing of supplies of energy, the development of key technologies, promoting innovations in nature conservancy and species protection or in international development cooperation (Coenen and Grunwald 2003; Chambers 1994). In practice, however, sustainability has been limited to a number of individual sectors, e.g. ensuring a viable pension system over the long term or maintaining economic competitiveness. There has not been enough systematic, integrative study of the three dimensions of sustainability. And the expansion of means of participation through new methods to a number of political levels – local, regional, national and international – and in diverse social spheres – political, economic, academic and educational – is still to a great extent selective and little institutionalised despite growing academic discussion about participatory governance (Delli Carpini 2004; Creighton 2005).

Behind both concepts there are far-reaching ideas, concepts and approaches for social modernization and transformation processes. It is unsurprising that a number of different interpretations and expectations meet in these fundamental perspectives. However there is still a broadly shared perception of the problem. The dynamic of social and biophysical changes – driven by globalisation and global environmental changes – requires new forms of communication in order to build collective opinion and decision-making processes (participation) as well as to create a more conscious orientation towards interdependent and temporal-spatially disassociated effects (sustainability).

Over the past 10 years in a number of different areas – from politics to economics to the educational system – there has been an increase in social activities concerning sustainable development. In spite of this development on both global and local levels of policy – and it should not be underestimated – changing the on-going non-sustainable development dynamic, under real-world conditions of power and interest relations, is a Herculean task (Steffen et al. 2004). Collective development and decision-making processes become even more difficult given the limits of knowledge regarding forecasting, risk, simulation and scenario and the accompanying uncertainties in diagnosing problems.

When considering the relationship between sustainability and participation and faced with cognitive uncertainty and normative ambivalence, it is clear that participation and participatory approaches need to be further developed if we are to improve anticipative knowledge communication and decision-making. This would allow a reduction of risky failures, in particular environmental ones, and the exploration of possibilities for sustainable development. Besides the use of participation methods in local Agenda 21 processes, there have been manifold 'experiments' with participatory approaches. Even innovative approaches developing and testing sustainability oriented participation methods have been explored. In the context of sustainability research, citizen participation has been used to diagnose problems and evaluate possible courses of action (Kasemir et al. 2003). This demonstrates that participation methods can make important contributions to the rationalisation of sustainability discourse and release creative potential. These impulses stimulate critical social learning, which is essential for the necessary transformation processes towards a sustainable society (Siebenhüner and Arnold 2007).

Building on this involves evaluating the legitimacy, effectiveness and efficiency of forms of participation for sustainability as well as developing innovative methodological designs – also regarding new media and communication technologies. In the differentiated, pluralistic and transnational (global) society of today, our view should not be limited to the political space and concrete decision-making processes. We need to widen the scope to analyse and develop participation and cooperation in other social contexts such as education, economics and science. The mass media also play an important role in structuring content in the public communication arena and by broadcasting information create the conditions for the participation of larger numbers of the population and of a greater variety of different social actors. In addition to the analysis and development of sustainability-oriented participation methods on a micro-sociological level, there is a need to identify the possibilities and limits of an institutional integration of participative activities on a meso-level and to observe social-material sustainability effects on the macro-level. Granted the necessity of continuing the just begun cultural evolution towards a participation, cooperation and sustainability oriented society, there is still much work to be done in research and development in the participative sustainability communication field. More analysis and more impulses are needed to supplement the dominant logic of hierarchically-based knowledge transfer and decision-making with a new logic based on functionally specific participative and cooperative knowledge discourse and decision-making. However, we do not have to start from zero; especially in democratic societies, sustainability-oriented participatory approaches can be developed out of existing cultures of participation.

The Requirements of Participation in Complex Societies

The discussion about participation, or more precisely political participation, is not of course a new one. The history of democracy as a form of social (self-) organization can be viewed as the history of increasing possibilities for more and more people to participate in collective processes of opinion-making, formulating political objectives and decision-making. In representative democracies conventional participation takes place mainly through elections, while unconventional participation on the other hand is found – and is guaranteed by the right to freedom of speech – in demonstrations and protests.

At the same time there have been, and still are, repeated demands for a further democratisation of democracy. Extended possibilities for participation and a greater involvement of citizens in collective decision-making processes is considered essential to reduce the possibility of alienation from the political system, or political disaffection, to find viable, socially acceptable and accepted solutions. In particular, in many countries since the 1960s there have been numerous academic debates and practical activities, all of which can be subsumed under the heading of the 'participative revolution'. These include radical and grass-roots democratic ideas in politics as well as the strengthening rights of worker co-determination in the economic sector (Rucht 1997).

These demands for democratisation, which were first put forward by new social movements, have led to an expansion of possibilities for participation by politically interested citizens and politically active workers. Especially in the case of environmentally relevant large-scale technological and infrastructure projects, these opportunities for political participation have become increasingly institutionalised (e.g. environmental impact tests). Participation was initially limited to an increased government obligation to make information publicly available as well as to more civil rights guaranteeing access to information and consultation. Since the beginning of the 1990s – triggered by the United Nations' Agenda 21 – there has been a world-wide wave of new interest in participation. The spectrum ranges from a greater participation of civil society actors (NGOs) in international conferences and negotiations to expanded rights to information for involved parties and citizens to the participation of interest groups and citizens in local Agenda 21 processes. As a result there is a continuing debate, especially in Western democracies, about the quality and quantity of social participation in collective decision-making and development processes (Dryzek 1994).

When discussing and implementing participative elements to expand representative democracy, two lines of argumentation are of central importance. First of all, there is an ethical-normative perspective, according to which it is in principle a good thing when as many people as possible are involved in the decisions that affect their lives. And second there is a functional-analytical viewpoint, according to which a representative political system can only deal with problems inadequately. Both lines of argumentation indicate that it is both necessary and desirable to involve a greater variety of actors as well as broader sectors of the population in the specific search, learning and development processes needed to adequately cope with the technological and social complexity of highly differentiated civil societies.

There are a large number of proponents, both nationally and internationally, of an expansion in participation, but there are also critical voices. From a perspective of consensus and conflict theory, one might for example ask to what extent participation is able to contribute to initiating social change. In development policy contexts, there are warnings that participation can be counterproductive by creating acceptance for existing structural inequalities rather than serving the empowerment and self-organization of the population (Cooke and Kothari 2001). From an administration theoretical perspective, it may be more important for there to be efficient 'public management' than to have the public participate in each and every issue that affects them (Dahl 1994). We should remember that, due to a lack of knowledge and competence, citizens are often unable to make an important contribution to solving many issues. These perspectives share a reference to the political elite model as well as a preference for bureaucratic-technocratic action together with scepticism towards the citizen and sovereign. Finally there are also demands for the democratic legitimation of participative procedures that – with the exception of petitions and citizen initiatives – do not involve the whole population

(Brown 2004). These and other critical voices are without doubt important if we are to avoid succumbing to naïve participation euphoria. However, in the discussion about participation, what is important throughout is not to lose sight of existing social inequalities in power and resources, to reflect on the efficiency and legitimation of participation in a representative democratic system and to make allowance for the possibility of citizens being confronted with excessive demands on their ability to contribute to sound decision-making.

In spite of this criticism it seems to be necessary, if we are to effectively deal with the existing social and technological complexity found in pluralistic knowledge societies, to institutionalise the expanded possibilities for participation (Heinrichs 2005). Of particular importance in this discussion are the dialogic-based participation methods that were conceptually developed, especially in the 1990s, and then put into practice in a number of different countries.

Participation Methods

The newer participation methods are significantly different from other conventional (elections, hearings) and unconventional (protests) possibilities in that they aim at being dialogic, discursive and deliberative. This means that, first of all, they should be structured, two-way communication and, secondly, conflicting arguments and claims should be related to each other in order to achieve consensus where there is dissensus and, thirdly, heterogeneous actors should develop solutions together in consultative processes. Participation processes thus aim at a systematic rationalisation of knowledge, value and interest pluralism in order to enable a cooperative process of understanding. There have been a number of different participation processes used since the 1970s to deal with environmental, technological and risk problems. According to Grunwald there are six central aspects (Grunwald 2002: 128f.):

- broadening the basis of knowledge for decision-making (supplementing expert specialist knowledge with local knowledge, experience and professional knowledge)
- broadening the basis of common values to increase the social stability of decisions
- making available more information to enable citizens to make informed judgements
- increasing social acceptance by including, and critically reflecting upon, a variety of claims and demands
- practicing conflict avoidance and management by using a cooperative search for objective solutions that can be supported by all
- developing an orientation to the common good by using rational discourse strategies to overcome particular interests of individuals

In contrast to the free political competition of opinions, to neo-corporatist negotiations and to hierarchical control instruments such as laws, economical and educational-informative approaches, participation methods offer a chance for the structured integration of diverging perspectives and the development of creative solutions to collective problems. In general we can define the newer participation methods as: "forums for exchange that are organized for the purpose of facilitating communication between government, citizens, stakeholders and interest groups, and businesses regarding a specific decision or problem" (Renn et al. 1995: 2).

The newer participation methods have in common that they offer a structured possibility for communication among heterogeneous groups of actors. At the same time there are also differences in the extent to which they depend on a particular political context and function. The following review shows the defining characteristics and uses of the most important participation models.

Mediation/Alternative Dispute Resolution (ADR)

Since the 1970s in the USA the so-called alternative dispute resolution procedure (ADR) has been developed and used so as to avoid long legal disputes and court cases. Mediation is the most common ADR method. It is used in both interpersonal and business conflicts, and especially in environmentally relevant projects regarding their location and infrastructure (Susskind and Fields 1996). In acute or threatening conflicts, 'neutral' and competent mediators provide moderation or mediation to support the conflict parties' search for solutions acceptable to both sides (win-win situation). By systematically revealing the particular interests and perspectives of each party, the method promises to find common ways of viewing a problem and finding a viable compromise as a solution. The willingness of both parties to negotiate constructively is a necessary condition for it to be used successfully. Mediation, which arose in the competitively organised political-legal system in the US, has been used in Germany since the 1990s to deal with on-going conflicts over the environment as well as to prevent their occurrence (Baugham 1995).

Stakeholder Dialogue

Companies, and especially those with environmentally sensitive production methods and products, such as the chemical industry, have been put under increasing public pressure since the 1970s by social movements and environmental organisations. Beyond environmental and legal requirements, they have to face increasing public discussion of the environmental and social compatibility and sustainability of their activities. With the method of stakeholder dialogue – i.e. the structured communication of all groups with a claim on the outcome of a particular activity – participative-cooperative elements have become part of the corporate policy of numerous (multinational) companies. The spectrum ranges from neighbourhood dialogues with the local population at production sites to industry-wide dialogs with NGOs.

Future analyses must examine to what extent shareholder dialogs actually fulfil the requirements of cooperative communication processes, reduce risks and stimulate sustainable production and consumption patterns, or whether they are more PR events for maintaining a corporation's reputation (Freeman 2007; Stoll-Kleemann and Welp 2006).

Round Table

The round table has become especially well known in eastern Germany since the German reunification. All of the relevant interest groups are represented at a table and in consensus-oriented negotiations develop options and solutions to complex problems. The moderated discussion should be discursive, i.e. the participants have the opportunity to articulate their claims as co-partners with equal rights and to work constructively on objective solutions. This procedure promises to coordinate the actions of heterogeneous actors towards a collectively shared goal. This method is suitable for optimising large-scale infrastructure location planning processes (e.g. finding a new location for a waste disposal site) as well as for reaching politically strategic decisions at, for example, so-called 'energy round tables' to plan new energy-saving projects (Knaus and Renn 1998).

Cooperative Discourse

Cooperative discourse is a sophisticated participation method in which elements from other approaches, such as mediation, round table and planning cells (see below), are combined in order to contribute to the solution of problems with high cognitive uncertainty and normative ambivalence. In this method the knowledge, value and interest pluralism common in many technological and environmental fields are analysed in a three-step procedure. First, a value tree analysis makes the values and preferences of stakeholders transparent and structures them by having the actors involved reveal their judgements about the proposed environmental policy measures. In a second step experts contribute their knowledge about possible courses of action. Since in many cases there is no certain knowledge available – especially concerning potential medium and long-term effects – experts from both sides of an issue are involved in order to disclose the complete spectrum of knowledge on the subject, from confirmed to unconfirmed to areas of ignorance. Finally, in a third step, knowledge of value dimensions and of the consequences to particular courses of action create a framework in which citizen forums can then discuss their concerns. By structurally integrating organised value representations through interest groups, expert knowledge and citizen preferences, it is possible to cooperatively resolve problems of technology, environment and risk (Knaus and Renn 1998).

Consensus Conference

Since the end of the 1980s in Scandinavian countries, the consensus conference has been developed and used as a citizens' panel with expert input. Representatively selected citizens discuss controversial topics – mainly in the area of innovative technologies. They make themselves familiar with the topic, also by using the opportunity to question experts, and then at the end of this discursive process they make a public statement presenting their conclusions to the decision-makers. Although this procedure is designed to achieve consensus, there is a possibility to voice a minority opinion. Consensus conferences, due to their discursively generated and informed citizens' opinion, have a rationalising input in publicly negotiated, often technological, controversies (Joss and Durant 1995).

Planning Cells/Citizens' Reports

Planning cells, which were developed in the 1970s by Dienel, involve a group of randomly chosen citizens to work on a clearly delineated communal, and often technological or environmental, problem. They familiarize themselves with the problem to be discussed, work with expert knowledge and then in a citizens' report develop recommendations for the planned project. This method was intended as a form of democratisation during the planning euphoria in the 1970s. Planning cells have since shown that citizens as laypeople are, after a short period of time, capable of producing informed reports that have creative solutions oriented toward the common good (Dienel and Renn 1995). Decision-makers are given an insight into the informed opinion of citizens about a specific planned project as well as innovative, publicly acceptable ideas that are adapted to the local conditions.

Future Workshop

The future workshop developed by Robert Jungk is neither designed for conflict resolution or preventative conflict avoidance nor for the consensus-oriented development of courses of action (Jungk and Müllert 1989). On the contrary this method is meant to produce creative ideas in response to the question: How do we want to live, work and act in the future? Group work is divided into three phases, with critical reflection of existing conditions in the first criticism phase, developing future scenarios in the second imagination or utopia phase, and finally finding ways to put such elements into practice in the third implementation phase. Especially in the first two phases, participants are encouraged to free themselves from any restrictions on using their imagination. It is not until the third phase that, with input from experts, the feasibility of the courses of action generated in the first two phases is critically evaluated. The future workshop is especially suited to developing creative and innovative solutions to problems accompanying new projects.

Scenario Workshop

The scenario method was originally developed for strategic planning in military and economic contexts as a supplement to forecasting instruments such as trend extrapolation or simulation (Reibnitz 1987). Starting with the insight that the future is in principle undetermined and that knowledge about future developments is always incomplete and uncertain, the scenario method aims at creating a space for possibilities about imaginable futures. Strategic courses of action are then developed to optimise the chances and risks for each scenario (Reibnitz 1987: 15–26). This method develops anticipatory knowledge and preventive options of action. While in the majority of cases this method has been used with non-lay professionals and experts, recently there have been successful examples of scenario workshops using citizens as a method of participative forecasting (Niewöhner et al. 2004). Experts are given an opportunity to contribute their knowledge in these processes. In contrast to future workshops, scenario workshops are more analytic and are not normatively oriented. In the first place this participation process is not about desirable futures, but about identifying imaginable futures. This involves integrating various sets of knowledge and experience, reflecting critically on values and preferences in discussion and then developing together with other participants in a workshop potential courses of action. An important side effect in scenario workshops is social learning, which is stimulated through the structured discussions about possible future developments.

eParticipation

eParticipation is not an independent method but rather a new information and communication technology that creates greater access to citizen participation (Fuchs and Kastenholz 2002). Complex information, such as blueprints, maps or any type of graphical representations, can be easily communicated. And moderated discussion forums offer virtual communication rooms that facilitate discussion with experts and the exchange of opinions. eParticipation is primarily suitable as a supplement to other methods of public participation, which for organizational reasons only permit a restricted number of individuals to enter into face-to-face communication. A limiting factor of this method is the socially unequal access to the internet.

The participatory methods outlined above and comparable approaches have been used over the past years in many Western democracies as well as in emerging and developing countries, especially in procedures assessing the consequences of technology, risk assessment, environmental conflicts and planning projects with regard to sustainable development. Even in non-democratic countries such as China experiments with participatory procedures can be observed (Horsley 2009). Due to the wide-spread use of specific participatory methods and approaches to sustainable development in different sectors such as stakeholder dialogue in business and policy making, citizen participation in policy-making, consumer participation in business, transdisciplinary projects in science or interactive communication in new media, it is difficult to assess the expansion of participatory culture and its substantial effects over the past decades. Nevertheless there is a growing interest in gaining better knowledge about participation and its effects for sustainable development. On the one hand innumerable single case studies around the world evaluating specific participation processes and their outcomes show under which circumstances participation can make a difference. The Austrian Ministry of Agriculture, Forestry, Environment and Water Management, for example, has set up a database for participatory activities in the European Union (http://www.partizipation.at/index. php?kontakt). Looking at the single case studies it becomes clear that if there is transparency concerning the scope of deliberation and participation as well as professional conduct in the participation process at hand then participative procedures have the potential to contribute positively to the development of sustainable solutions. Comparative approaches analysing multiple cases from a meta-perspective have been shown to be useful for improving understanding of participatory processes. Newig and Fritsch for example conducted a meta-study comparing 42 single case studies in participatory environmental governance. Their key insights are that environmental pre-attitude and face-to-face communication have a positive effect on environmentally positive outcomes and that context and process factors, which are however increasingly absent from generalising meta-studies, are central for the understanding of specific participatory processes. They conclude that, depending on the individual context and process, participation *can* have a positive influence on environmental governance. Even though this study focused only on the environmental outcomes of participation and did not look at the co-optimization of social, economic and environmental dynamics, this method provides a useful approach for analyzing the wider effects of participation. After 20 years of experiments with informal participation methods in a broad range of different settings, there is certainly a need for more studies evaluating the cumulated macro-effects of participation processes for sustainable development, going beyond single case studies and single dimension studies focussing only on environmental outcomes.

In Germany these participatory methods have not become institutionalized, routine instruments of participative policy making but they have become widespread. And, despite debates about effectiveness and legitimation, the experiences that have already been made are generally encouraging. Their dialogic, discursive and deliberative character has the potential to stimulate social learning, build social capital and rationalise conflicts, generate creative and viable solutions and constructively deal with existing social and technological problems. The potential of participation methods needs to be better made use of in the future for the integrative resolution of sustainability problems (Siebenhüner and Heinrichs 2010).

Over the past two decades experiments with participatory approaches as interactive procedures have contributed new ways to the societal handling of – socially and factually – complex problems, especially in the context of environmental, technological as well as urban and regional planning. Regarding the broader context of sustainable development, existing and novel participatory approaches have a great potential for identifying and shaping sustainability solutions – and in doing so limiting social conflicts around sustainability challenges. Along with persuasive campaigning, educational and mass-mediated means of sustainability communication, the general idea and the concrete procedures of participation have much to offer in helping societies find their way to sustainable development.

References

Baugham, M. (1995). Mediation. In O. Renn, T. Webler, & P. Wiedemann (Eds.), *Fairness and competence in citizen participation* (pp. 117–140). Dordrecht: Springer.
Brown, M. (2004, July). *Citizen panels and the concept of representation*. Working paper presented at the annual workshop of the "Science and Democracy Network", Boston.
Chambers, R. (1994). Participatory rural appraisal (PRA): Analysis of experience. *World Development, 22*(9), 1253–68.
Coenen, R., & Grunwald, A. (Eds.). (2003). *Nachhaltigkeitsprobleme in Deutschland – analyse und Lösungsstrategien*. Berlin: Edition Sigma.
Cooke, B., & Kothari, U. (2001). *Participation: The new tyranny?* London: Zed Books.
Creighton, J. L. (2005). *The public participation handbook: Making better decisions through citizen involvement*. San Francisco: Jossey-Bass.
Dahl, R. A. (1994). A democratic dilemma: System effectiveness versus citizen participation. *Political Science Quarterly, 109*(1), 23–34.
Delli Carpini, M. X. (2004). Public deliberation, discursive participation, and citizen engagement. A review of the empirical literature. *Annual Review of Political Science, 7*, 315–344.
Dienel, P. C., & Renn, O. (1995). Planning cells: A gate to "Fractal" mediation. In O. Renn, T. Webler, & P. Wiedemann (Eds.), *Fairness and competence in citizen participation* (pp. 117–140). Dordrecht: Springer.
Dryzek, J. S. (1994). *Discursive democracy: Politics, policy, and political science*. New York: Cambridge University Press.
Freeman, E. R. (2007). *Managing for stakeholders: Survival reputation and success*. New Haven: Yale University Press.
Fuchs, G., & Kastenholz, H. (2002). E-democracy: Erwartungen der Bürger und erste Realisierungen. Ein Werkstattbericht. *Technikfolgenabschätzung, 11*(3/4), 82–91.
Grunwald, A. (2002). *Technikfolgenabschätzung – eine Einführung*. Berlin: Edition Sigma.
Heinrichs, H. (2005). Advisory systems in pluralistic knowledge societies: A criteria-based typology to assess and optimize environmental policy advice. In P. Weingart & S. Maasen (Eds.), *Democratization of expertise? Exploring novel forms of scientific advice in political decision-making* (pp. 41–61). Dordrecht: Springer.
Horsley, J. P. (2009). Public participation in the people's republic: developing a more participatory governance model in China. Retrieved July 30, 2010, from Yale Law School Website: www.law.yale.edu/documents/pdf/Intellectual_Life/CL-PP-PP_in_the__PRC_FINAL_91609.pdf.
Joss, S., & Durant, J. (1995). *Public participation in science. The role of consensus conferences in Europe*. London: Science Museum.
Jungk, R., & Müllert, N. (1989). *Zukunftswerkstätten*. Munich: Heyne.

Kasemir, B., Jäger, J., Jaeger, C. C., & Gardner, M. T. (Eds.). (2003). *Public participation in sustainability science*. Cambridge: Cambridge University Press.

Knaus, A., & Renn, O. (1998). *Auf dem Weg zum Gipfel. Unterwegs in eine nachhaltige Zukunft*. Marburg: Metropolis.

Niewöhner, J., Wiedemann, P., Karger, C., Schicktanz, S., & Tannert, C. (2004). Participatory prognostics in Germany - developing lay scenarios for the relationship between biomedicine and the economy in 2014. *Technological Forecasting and Social Change, 72*(2), 195–211.

Renn, O., Webler, T., & Wiedemann, P. (Eds.). (1995). *Fairness and competence in citizen participation. Evaluating models for environmental discourse*. Dordrecht: Kluwer.

Rucht, D. (1997). Soziale Bewegungen als demokratische Produktivkraft. In A. Klein & R. Schmalz-Bruns (Eds.), *Politische Beteiligung und Bürgerengagement in Deutschland. Möglichkeiten und Grenzen* (pp. 382–403). Bonn: Bundeszentrale für politische Bildung.

Siebenhüner, B., & Arnold, M. (2007). Organizational learning to manage sustainable development. *Business Strategy and the Environment, 16*, 339–353.

Siebenhüner, B., & Heinrichs, H. (2010). Knowledge and social learning for sustainable development. In M. Gross & H. Heinrichs (Eds.), *Environmental sociology. European perspectives and interdisciplinary challenges* (pp. 185–200). Dordrecht: Springer.

Steffen, W. L., Sanderson, A., Tyson, P., Jäger, J., Matson, P. A., Moore, N., III, Oldfield, F., Richardson, K., Schellnhuber, H. J., Turner, B. L., II, & Wasson, R. J. (Eds.). (2004). *Global change and the earth system: A planet under pressure*. New York: Springer.

Stoll-Kleemann, S., & Welp, M. (2006). *Stakeholder dialogues in natural resources management*. Berlin: Springer.

Susskind, L. E., & Fields, P. (1996). *Dealing with an angry public: The mutual gains approach to resolving disputes*. New York: The Free Press.

von Reibnitz, U. (1987). *Szenarien – Optionen für die Zukunft*. Hamburg: Mc Graw-Hill.

Index

A
Action Plan on Sustainable Consumption and Production (SCP), 142, 143
Agenda 21, 4, 97, 190, 192
Agenda-setting, 83, 84, 122
Agrobiodiversity, 132, 135
Alternative dispute resolution, 194
Ambiguity, 5, 8, 45, 47, 115, 181
Approach
 inside-out, 153, 166, 167
 integrative, 27–35
 outside-in, 153, 166, 167
 participatory, 126, 142, 190, 191, 199
 phenomenological, 57
 post-structuralist, 57
Assessment, 16, 20, 30, 31, 34, 56, 83, 96, 163–164, 168, 198
Assurance, 163–165
Attitude-behaviour gap, 145
Auditing, 163–164, 168
Autopoiesis, 55, 109, 111
Awareness, 5–8, 10, 34, 47, 65, 74, 79, 87, 89, 98, 130, 137, 145

B
Basic needs, 17, 143–145
Behaviour patterns, 46, 69–73
Benchmarking, 154, 166
Biodiversity, 4, 10, 129–138
Biological diversity, 129, 130, 133–136
Biosphere, 19, 137
Broad interdisciplinarity, 44

Broadcast, 73, 84, 95, 121, 148, 191
Brundtland
 commission, 4
 report, 5, 13

C
Campaigns, 33, 62, 64, 98, 125, 141–144, 148, 199
Civil society, 9, 11, 23, 126, 127, 192
Climate change, 4, 8, 15, 16, 28, 31, 33, 35, 61, 72, 74, 119–127, 133, 142, 181
Club of Rome, 3
Cognitions, 71, 75, 109, 112–113
Collaboration, 39–47
Commitment, 14, 75, 89, 94, 95, 145, 159, 177
Common ground, 47, 48
Communication
 climate change, 35, 119–126
 computer-supported, 175–177
 environmental, 28–31, 34, 35
 global, 4, 79, 80, 85, 86
 instruments, 75
 interpersonal, 72, 123
 model, 33, 75
 modes of, 123, 126, 142
 paradox of, 178
 personal, 121
 risk, 28, 30–32, 34, 35, 124
 sciences, 27, 28, 32–35, 112, 147
 theory, 6, 7, 89–96, 125
 thread, 80
Communicative process, 14, 89

J. Godemann and G. Michelsen (eds.), *Sustainability Communication: Interdisciplinary Perspectives and Theoretical Foundations*, DOI 10.1007/978-94-007-1697-1,
© Springer Science+Business Media B.V. 2011

Competence(s)
  development models, 103
  key, 102
  models, 102, 104
  structural models, 103
Complexity, 7, 19, 23, 29, 30, 34, 35, 43, 44, 71, 91, 96, 99, 110, 114, 119, 123, 159, 161, 178–180, 189, 192, 193
Computer, 4, 72, 85, 173–185
Concept of correspondence, 174
Connectivism, 85
Conservation psychology, 136
Constructivism
  neuro-biological, 111
  radical, 113
Consumption
  culture, 147
  strategic, 142
  sustainable, 65, 141–148
Conversation support systems, 176, 182, 184, 185
Cooperation, 31, 39, 61, 70, 91, 95, 104, 134, 136, 148, 181, 190, 191
Council on Environmental, 4
Critical-constructive didactics, 105
Cross-media concepts, 148
Cultural
  communication, 83
  context, 5
  diversity, 129, 135–136, 138
  evolution, 191
  identity, 83

D
Decision
  making, 15, 22, 24, 31, 33, 45, 46, 71, 72, 95, 98, 99, 104, 120, 157, 173, 175, 177, 178, 184, 189–193
  support systems, 175, 184, 185
Democracy, 32, 82, 191, 192
DeSeCo competency concept, 102
Digital
  divide, 4
  natives, 85
Disciplinarity
  informed, 41, 42
Disciplines, 6, 29, 32, 34, 39–49, 57, 69, 70, 111, 112
Discourse
  elite, 126, 127
  ethics, 15, 16
  global, 79, 127
  political, 123
  scientific, 6, 123
  theory, 56, 57
Disenchantment, 33

E
Earth summit, 28
Ecological
  danger, 4
  discourse, 28, 89
  effectiveness, 156
  systems, 20, 111
Economic
  effectiveness, 156
Ecotainment, 141, 147, 148
Education
  categorial, 106
  for sustainable development (ESD), 10, 97–106
  general, 105
  measurability of, 100–101
  monitoring, 99, 100
Efficiency, 9, 16, 73, 144, 154, 178, 179, 191, 193
Emotion, 74, 75, 112–113, 147
Empowerment, 9, 23, 30, 189–199
Environmental
  action, 72–73
  awareness, 72–74
  communication, 28–31, 34, 35
  ethics, 16, 20, 92
  NGO, 124, 125
  policy, 11, 23, 29–31, 195
  psychology, 9, 69, 74
  reporting, 34, 156, 166
  sociology, 56
eParticipation, 10, 197–199
Epistemology, 6, 110–112
Ethics, 15, 16, 22
European Union Eco-Management and Audit Scheme (EMAS), 156, 165
Evaluation, 10, 41, 44, 46, 70–72, 75, 95, 99, 106, 174, 175

F
Face-to-face interactions, 70
Feedback, 75, 76, 146
Flexible intelligence, 111
Framework
  theoretical, 3, 6, 9, 13, 14, 16, 85, 106
Frankfurt School, 82

Index

Functional differentiation, 96
Future workshops, 10, 196–197

G

G3 guidelines, 160, 161, 163
German Advisory Council of Global Change (WBGU), 8, 21, 30, 47, 48, 183
Gestaltungskompetenz, 103, 104, 106
Globalisation, 3, 4, 61, 83, 190
Governance, 33, 124–126, 164, 190, 198
Governing, 56–58, 92, 120
Group work, 194
Guideline(s), 95, 104, 148, 160–161, 163, 166, 168

H

Human
    ecology, 70
    nature interrelationships, 131

I

Idée directrice, 57
Images, 8, 24, 35, 59, 72, 94, 113, 125, 147, 165, 178, 180, 184
Indeterminacy, 7
Indicators, 8, 16, 72, 99–101, 104–106, 122, 125, 162, 166
Indigenous people, 136
Individuality, 17
Information
    asymmetry, 159
Institutional practices, 8, 24, 55–58, 62
Instrument
    persuasive, 11, 29
    political, 11, 126
Intentions, 5, 7, 75, 90, 91, 94, 120, 124, 148
Interdisciplinarity
    conceptual, 41, 42
    synthetic, 41, 42
Intervention(s), 10, 34, 60, 73–76, 99, 110, 124, 130
IPCC report on climate change, 61, 121, 122
Issue attention cycle, 121

J

Justice
    intergenerational, 16, 73, 94, 189
    intragenerational, 6, 14, 15, 94

K

Knowledge
    declarative, 103, 106
    domain-specific, 101
    gap between knowledge and action, 73
    integration of, 41–43, 48
    production, 5, 57

L

Language-action, 173, 183, 184
Languages, 7, 14, 18, 22, 32, 43, 72, 79, 90, 91, 99, 102, 109, 113–114, 125, 135, 174–176, 178, 180, 182–184
Laypersons, 124, 125
Legitimacy, 32, 97, 105, 106, 154, 191
Life cycle assessment (LCA), 174, 175, 178, 180, 183, 184
Lifestyle of Health and Sustainability (LOHAS), 63
Lifestyles, 5, 8, 9, 24, 55, 62–65, 70, 72, 76, 97, 133, 141, 143
The limits of growth, 4, 174

M

Management
    information systems (MIS), 173, 174
    rules, 21
Media
    communication, 75, 79, 168
    cross-media concepts, 148
    discourse, 122, 123
    mass, 7, 23, 29, 34, 35, 57, 61, 62, 65, 72, 83, 85, 89, 92, 94, 95, 120, 121, 123, 191
    new, 79, 80, 82, 83, 85–87, 142, 191, 198
    print, 84
    science, 80, 84, 148
    technology, 93
    theory, 6, 7, 79–87
Medialisation, 7, 94–95
Mediation, 10, 76, 194, 195
Mental models, 46
Messages, 8, 80, 82, 83, 124, 125, 131, 143, 146–148, 177
Metaphor, 40, 48, 59, 125, 178, 180, 183
Methods, 3, 8–10, 40, 41, 43, 44, 46, 48, 74, 98, 100, 102, 126, 134, 174, 181, 184, 190, 191, 193–199
Milieu, 8, 55, 63, 64, 70, 143
Mode 1 / Mode 2, 5, 10, 15, 41, 56, 76, 120, 123, 126, 142

Moral, 16–18, 20, 22, 24, 29, 43, 95, 96, 104, 112, 145, 146, 157
Mutual learning, 47

N
Natural capital, 15, 18–23
Nature conservation, 14, 20, 137
Non-sustainability, 3, 9, 15, 48, 65, 69, 72, 73, 98, 184
Normalisation, 7, 89, 94
Norms, 6, 7, 55, 58, 63, 64, 69, 74, 76, 91, 124, 146, 155, 178, 182

P
Paradigms, 5, 44, 80–83, 109, 111, 181
Participation
    methods, 190, 191, 193–199
    of civil society, 127, 192
Perceptions, 5–8, 28, 30, 35, 62, 69–72, 80, 81, 83, 84, 87, 94, 112, 113, 123, 126, 142, 159, 166, 167, 174, 190
Personal life transitions, 147
Philosophy, 14, 21, 40
Planning cells, 195, 196
Popularisation, 33, 92, 95
Precautionary principle, 20, 31
Problem-solving, 5, 10, 40, 44–47, 70, 72
Psychology, 6, 9, 47, 63, 69, 73, 74, 76, 136, 145
Public
    attention, 35, 63, 123, 127
    understanding of science, 33, 125

R
Ranking, 84, 93, 165–166
Rating, 165–168
Receiver, 6, 90, 124
Recipient(s), 75, 82–84, 124
Reflexivity, 7, 44, 45, 89, 93
Regulations, 30, 58, 84, 123, 126, 142, 160–163
Report(ing)
    citizen(s), 196
    corporate sustainability, 153–168
    environmental, 34, 156–158, 162, 166, 184
    extra-financial, 153, 155, 156, 159, 163
    integrated (business), 158
    internet-supported sustainability, 165
    non-financial, 154, 162
    online sustainability, 164, 165

Reputation, 154, 168, 195
Resilience, 16
Responsibility, 4, 5, 16, 44, 62, 74, 87, 112, 145, 159
Rio Conference, 59
Risk
    communication, 28, 30–32, 34, 35, 124
    perception, 8
    research, 8
Round tables, 10, 195

S
Scenario workshop, 197
Science
    communication, 27, 28, 32–35, 112, 147
    new modes of, 120
    system, 32, 109, 110
Self-reference, 109, 111
Sender(s), 6, 90, 124
Shared reality, 46
Social
    discourse, 3, 6, 7, 79–81
    learning, 39, 120, 191, 197, 198
    marketing, 9, 35, 76
    networks, 7, 9, 74, 142, 146, 173, 179, 183
    reports, 156–158
Socio-effectiveness, 156
Sociology, 6, 47, 56, 63, 65, 121
Stakeholder(s)
    and theme-specific reports, 157
    dialogue, 165, 166, 168, 194–195, 198
    involvement of, 120, 158
Sufficiency, 16
Sustainability
    behaviour, 69, 72–75
    corporate, 153–168, 175, 177, 179, 180, 184, 185
    discourse, 7, 14, 28, 55, 60, 62–65, 89–96, 147, 190
    reports(ing), 153–168
    research, 5, 39–40, 85, 165, 190
    strong, 5, 13–24
    weak, 5, 16, 19, 20
Sustainable
    consumption, 65, 141–148
    development, 3–6, 9, 11, 13, 14, 16, 23, 24, 30, 33, 34, 40, 48, 58–62, 65, 69–71, 73, 75, 76, 79, 86, 89, 91, 92, 94, 95, 97–106, 110, 113, 127, 129, 130, 138, 143, 147, 155, 157, 158, 167, 168, 173, 174, 177, 179–183, 185, 189–199
    science, 34, 40

Symbolic interactionism, 6, 8, 113, 147
Symbols, 57, 90, 174, 176
Syndrome concept, 47, 183
System
  theory, 6, 81–82, 90, 109–111
Systemic thinking, 109–111

T
Target groups, 9, 32, 34, 43, 64, 75, 101, 146–148, 156, 159, 175
Technology, 31, 32, 80, 89, 93, 105, 111, 143–145, 195, 197, 198
Television, 31, 73, 84, 85, 121, 147
Theory
  action, 56
  of cognition, 112
  of communicative action, 173
  of justice, 16
  of structuration, 56
  of sustainability, 14, 15, 143
  rational choice, 63
  speech act, 175, 185
  structuration, 8
  systems, 6, 81–82, 90, 109–111
Three-pillar model, 15, 22, 61

Transdisciplinarity
  research, 40, 42
Transformation, 57, 123, 174, 179, 182, 183, 190, 191

U
Uncertainty, 7, 29, 30, 34, 35, 45, 71, 99, 115, 119, 190, 195
United Nations Conference on Environment and Development, 4

V
Value, 6–9, 14, 18–20, 23, 24, 32, 33, 46, 63, 64, 69, 74, 80, 90, 91, 93–95, 106, 109, 120, 124, 129, 132, 137, 142–144, 154, 157, 164, 165, 174, 178, 179, 182, 184, 193, 195, 197
Vested interests, 126

W
Waldsterben, 70
Web 2.0, 7, 9, 79, 85, 86, 142, 182, 183
Wilderness, 134–135

Printed by Publishers' Graphics LLC